ENGINEERING SCIENCE
Third Edition

Engineering Science

THIRD EDITION

EDWARD HUGHES

D.Sc.(Eng.), Ph.D., C.Eng., F.I.E.E.
Fellow of the Heriot-Watt University Edinburgh
Formerly Vice-Principal and Head of the Engineering Department
Brighton College of Technology

and

CHRISTOPHER HUGHES

B.Sc.(Eng.), C.Eng., M.I.Mech. E.
Senior Lecturer in Mechanical Engineering
North Gloucestershire College of Technology,
Cheltenham

Longman Scientific & Technical

Longman Scientific & Technical,
Longman Group UK Limited,
Longman House, Burnt Mill, Harlow,
Essex CM20 2JE, England
and Associated Companies throughout the world.

First published 1970
Second Impression 1971
Thrid Impression 1972
Fourth Impression 1974
Second Edition 1977
Second Impression 1978
Third Impression 1979
Fourth Impression 1981
Third Edition 1987
Second Impression 1990

British Library Cataloguing in Publication Data
Hughes, Edward, *b. 1888*–
   Engineering science.—3rd ed.
   1. Physics
   I. Title II. Hughes, Christopher III. Hughes, Edward, *b. 1888.*
   Engineering science in SI units
   530′.0246   QC21.2

   ISBN 0-582-41380-X

Set in 10/12pt "*Monophoto*" Times

Printed in Malaysia
by Percetakan Jiwabaru Sdn. Bhd., Bangi, Selangor Darul Ehsan

Contents

Preface to third edition

The third edition has involved some reorganization of text and the introduction of new material to take account of changes occurring in syllabuses and the types of courses now being offered. Thus, account has been taken of BTEC National Certificate and Diploma course syllabuses in Engineering (the book being primarily levels I and II), the City and Guilds Basic Engineering Competences (201) Science Background to Technology, the Further Education Unit publication *Core Competences for Engineers* (July 1984) and the Certificate of Pre-Vocational Education (CPVE) preparatory module for Physical Science. The aim has been to produce a text which covers comprehensively basic engineering science.

W.B.

Symbols and abbreviations

Based on British Standard 3763:1970, and *The International System of Units*, 1970 (HMSO).

Notes on the use of symbols and abbreviations

1. A unit symbol is the same for the singular and the plural: for example, 10 kg, 5 V.

2. Full point should be omitted after a unit symbol.

3. Full point should be used in a multi-word abbreviation: for example, s.t.p., e.m.f.

4. In a compound-unit symbol, the product of two units is preferably indicated by a dot, especially in manuscript. The dot may be dispensed with when there is no risk of confusion with another unit symbol: for example, N·m or N m, but not mN.

   A solidus (/) denotes division: for example, m/s, J/kg.

5. A unit symbol should be used only after a numerical value: for example, *m* kilograms, 5 kg; *I* amperes, 10 A.

6. A hyphen is inserted between the numerical value and the unit when the combination is used adjectivally: for example, a 2-metre (or 2-m) rod; a 240-volt (or 240-V) supply.

7. The abbreviations 'a.c.' and 'd.c.' should only be used adjectivally: for example, a.c. circuit, d.c. motor.

1. GENERAL

Quantity	Quantity symbol	Unit	Unit symbol
Acceleration, linear	a, f	metre per second squared	m/s²
Acceleration, local gravitational	g	metre per second squared	m/s²
Acceleration, angular	α	radian per second squared	rad/s²

Quantity	Quantity symbol	Unit	Unit symbol
Angle, plane	α, θ	radian	rad
		degree	°
Density	ρ	kilogram per cubic metre	kg/m³
Density, relative	d	–	–
Distance along path	s	metre	m
Efficiency	η	–	–
Diameter	d	metre	m
Energy	W	joule	J
		kilojoule	kJ
		megajoule	MJ
		watt hour	W h
		kilowatt hour	kW h
Force	F	newton	N
		kilonewton	kN
Length	l	metre	m
		kilometre	km
Mass	m	kilogram	kg
		tonne (= 1000 kg)	t
Power	P	watt	W
		kilowatt	kW
Pressure	p	newton per square metre	N/m²
		pascal (= 1 N/m²)	Pa
Time	t	second	s
		minute	min
		hour	h
Torque	T	newton metre	N m
Velocity, linear	u, v	metre per second	m/s
		kilometre per hour	km/h
Velocity, angular	ω	radian per second	rad/s
Volume	V	cubic metre	m³
		litre (= 0·001 m³)	l
Weight	W	newton	N
Work	W	joule	J

2. MECHANICS AND THERMODYNAMICS

Coefficient of friction	μ		–
Energy, kinetic	T	joule	J
Energy, potential	V	joule	J
Momentum	p	kilogram metre per second	kg m/s
Young's modulus	E	newton per square metre or pascal	N/m² Pa
		giganewton per square metre or gigapascal	GN/m² GPa
Coefficient of linear expansion	α	per degree Celsius	/°C

Quantity	Quantity symbol	Unit	Unit symbol
Coefficient of cubic expansion	γ	per degree Celsius	/°C
Heat, latent	L	joule	J
		kilojoule	kJ
Heat, specific latent	l	joule per kilogram	J/kg
		kilojoule per kilogram	kJ/kg
Heat, quantity of	Q	joule	J
Specific heat capacity	c	joule per kilogram degree Celsius	J/kg °C
Temperature, thermodynamic	T	kelvin	K
Temperature, Celsius	θ, t^*	degree Celsius	°C
Temperature interval	–	kelvin, degree Celsius	K, °C

3. ELECTROMAGNETISM

Quantity	Quantity symbol	Unit	Unit symbol
Charge or Quantity of electricity	Q	coulomb	C
Conductance	G	siemens	S
Conductivity	σ	siemens per metre	S/m
Current:			
Steady value	I	ampere	A
		milliampere	mA
		microampere	μA
Instantaneous value	i		
Current density	J	ampere per square metre	A/m²
Difference of potential:			
Steady value	V	volt	V
Instantaneous value	v		
Electromotive force:			
Steady value	E	volt	V
Instantaneous value	e		
Frequency	f	hertz	Hz
Inductance, self	L	henry (plural, henrys)	H
Inductance, mutual	M	henry (plural, henrys)	H
Magnetic flux	Φ	weber	Wb
Magnetic flux density	B	tesla	T
Resistance	R	ohm	Ω
		microhm	μΩ
		megohm	MΩ
Resistivity	ρ	ohm metre	Ω m
		microhm metre	μΩ m

* In B.S. 1991: Part 1, 1967, θ and t are given as alternative symbols for temperature, with θ as the preferred symbol. Most standard textbooks on Thermodynamics use t; consequently, t has been adopted in this book.

Temperature *difference* in degrees Celsius is the same as in kelvins, i.e. change of temperature of 1°C = change of temperature of 1 K.

4. PREFIXES DENOTING DECIMAL MULTIPLES or SUB-MULTIPLES

Value	Prefix	Symbol
10^{12}	tera	T
10^9	giga*	G
10^6	mega	M
10^3	kilo	k
10^2†	hecto	h
10^1†	deca	da
10^{-1}†	deci	d
10^{-2}†	centi	c
10^{-3}	milli	m
10^{-6}	micro	μ
10^{-9}	nano	n
10^{-12}	pico	p

There should be no space or hyphen between the prefix and the name of the unit which it qualifies. Similarly, there should be no space or hyphen between the symbols for the prefix and the unit. For example: kilogram, kg; microsecond, μs. A double prefix such as kilokilogram should never be used.

5. GREEK LETTERS USED AS SYMBOLS IN THIS BOOK

Letter	Capital	Lower case
Alpha	–	α (angle, coefficient of linear expansion, temperature coefficient of resistance)
Beta	–	β (coefficient of superficial expansion)
Gamma	–	γ (coefficient of cubic expansion)
Eta	–	η (efficiency)
Theta	–	θ (angle, temperature)
Mu	–	μ (micro, coefficient of friction)
Pi	–	π (circumference/diameter)
Rho	–	ρ (density, resistivity)
Sigma	–	σ (conductivity)
Phi	Φ (magnetic flux)	φ (angle)
Omega	Ω (ohm)	ω (angular velocity)

* *Giga* is derived from a Greek word meaning giant and dictionaries state that its pronunciation should be the same as 'giga' in gigantic.

† Powers which are a multiple of 3 are generally preferred, but because of the large value that may result in the case of some derived units, such as volume, these smaller multiples are permissible, but their use should be limited as far as possible.

CHAPTER 1

Units

1.1 The International System of Units (SI)

The International System of Units, known as SI in every language, derives all the units used in the various technologies from the following *seven* base units:

Quantity	Unit	Symbol
length	metre	m
mass	kilogram	kg
time	second	s
electric current	ampere	A
temperature	kelvin	K
luminous intensity	candela	cd
amount of substance	mole	mol

The candela and the mole are not dealt with in this book and will therefore not be referred to again.

The SI* is a *coherent* system of units, i.e. the product or quotient of any two unit quantities in the system is the unit of the resultant quantity. For example, unit area (= 1 square metre) results when unit length (= 1 metre) is multiplied by unit length,

or 1 square metre = 1 metre × 1 metre.

Further examples illustrating the coherent nature of SI are:

(*a*) unit velocity (= 1 m/s) results when unit length (= 1 m) is divided by unit time (= 1 s);

(*b*) unit force (= 1 newton) results when unit mass (= 1 kg) is multiplied by unit acceleration (= 1 m/s²).

---

* It is incorrect to speak of 'SI system'.

In practice, it is often convenient to use a multiple or sub-multiple of the SI base unit, but the choice should, in general, be confined to powers of ten* which are a multiple of ± 3, thereby effecting a considerable reduction in the number of multiples and submultiples,

e.g.   1 kilometre $= 10^3$ metres,

1 millimetre $= 10^{-3}$ metre,

1 micrometre‡ $= 10^{-6}$ metre.

For *communication* purposes it is usually convenient to express a quantity in terms of a multiple or submultiple of the SI base unit, thereby avoiding the use of very large or very small numbers. For example, it is convenient to express the calorific value of coal as, say, 30 megajoules per kilogram rather than 30 000 000 joules per kilogram. On the other hand, for *calculation* purposes and especially in equations, it is advisable to express the quantities in terms of the SI base units and to insert the unit symbol in brackets† after the numerical value. For example, when clockwise and anticlockwise moments are being equated, the values should be expressed in the following form:

$$1000 \text{ [N]} \times 0 \cdot 008 \text{ [m]} = 16 \text{ [N]} \times 0 \cdot 5 \text{ [m]}.$$

This precaution ensures that the two sides of the equation are dimensionally the same.

1.2 SI Base Units of Length, Mass and Time

(*a*) *Metre.* The metre is the SI unit of length and is defined as the length equal to 1 650 763·73 wavelengths of the orange line in the spectrum of an internationally-specified krypton discharge lamp.

---

* It is important that students should be familiar with the use of indices and with the multiplication and division of numbers expressed in the form 10^n. For example, 2 000 000 can more conveniently be written as 2×10^6, and an error of a nought is far more likely to be made in the extended than in the shorter form.

When multiplying, say, 10^6 by 10^3, we add the indices; and when dividing 10^6 by 10^3, we subtract the indices: thus,

$$10^6 \times 10^3 = 10^9 \quad \text{and} \quad 10^6 \div 10^3 = 10^3.$$

In general, $10^a \times 10^b = 10^{(a+b)}$ and $10^a \div 10^b = 10^{(a-b)}$

‡ The use of the term 'micron' for 'micrometre' is deprecated by CGPM.

† Brackets, [ ], are more satisfactory than parenthesis, ( ), as they indicate more definitely that, say, 10[m] represents 10 metres and not $10 \times m$.

This definition reproduces the metre with an accuracy of one part in a hundred million (10^8). The metre was formerly defined as the distance between two lines on a certain platinum-iridium bar at 0°C. This bar is kept at the International Bureau of Weights and Measures at Sèvres, near Paris, and is still used as a reference standard. Its exact length is periodically determined in terms of the wavelengths of radiation from the krypton lamp.

(*b*) *Kilogram*. The kilogram is the SI unit of mass and is defined as the mass of a platinum-iridium cylinder kept at Sèvres. The mass of another body can be compared with that of this standard cylinder and can be determined with an accuracy of one part in a hundred million (10^8) by means of a specially-constructed balance.

(*c*) *Second*. The second is the SI unit of time and is defined as the interval occupied by 9 192 631 770 cycles of radiation corresponding to the transition of the caesium-133 atom. This definition enables the fantastic precision of one part in ten thousand million (10^{10}) to be achieved. The second is approximately 1/86 400 of the mean solar day.

$$1 \text{ minute [min]} = 60 \text{ seconds [s]}$$
and
$$1 \text{ hour [h]} = 3600 \text{ seconds.}$$

The minute and the hour are not decimal multiples of the second and are therefore non-SI units, but they are so firmly established in practice that their use is likely to continue indefinitely.

The SI base units of electric current and of temperature respectively are discussed in the chapters relating to their applications (chapters 15 and 13 respectively).

The degree of precision in the determination of the fundamental units of length, mass and time is really beyond human comprehension and is possible only with the elaborate apparatus available at national laboratories such as the National Physical Laboratory in this country. No student should be expected to memorize the figures given above.

For normal practical purposes, sub-standards of *length* and *mass* are used. The exact value of these sub-standards are periodically checked against those of the standards kept at the national laboratories. Sub-standards of *time*, such as clocks and watches, can easily be checked against time signals transmitted from observatories or from broadcasting sources such as the B.B.C.

1.3 SI derived* units of area and volume

If the floor of a room is 6 m long and 4 m wide,

$$\text{area of floor} = 6 \text{ [m]} \times 4 \text{ [m]} = 24 \text{ m}^2.$$

It will be noted that the unit symbol for square metre is 'm²' and not 'sq. m.'.

If a room has a floor area of 24 m² and a height of 5 m,

$$\text{volume of room} = 24 \text{ [m}^2] \times 5 \text{ [m]} = 120 \text{ m}^3.$$

An alternative unit of volume, often used in the metric system, is the *litre* (symbol, l). In 1964, the General Conference of Weights and Measures decided that the litre should be used as a special name for 1000 cm³ (or 0·001 m³) and not as the volume of 1 kg of pure water at maximum density, as had previously been the practice. Precise measurement has shown that 1 kg of pure water at maximum density and under atmospheric pressure is 1·000 028 litres; but for most practical purposes, we can assume that 1 litre of water has a mass of 1 kg.

1.4 Density and relative density

The *density* of a body is defined as the *mass per unit volume* and is represented by the Greek letter ρ (rho). Thus, if a body has a mass of m kilograms and a volume of V cubic metres,

$$\text{density} = \rho = m/V \text{ kilograms/cubic metre.}$$

It was mentioned in section 1.3 that the mass of 1 litre of water is practically 1 kg.

$$\text{Since } 1 \text{ m}^3 = 1000 \text{ litres}$$
$$\therefore \quad \text{mass of 1 m}^3 \text{ of water} = 1000 \text{ kg}$$
$$\text{and density of water} = 1000 \text{ kg/m}^3.$$

The *relative density* (symbol, d) of a material is defined as the ratio:

$$\frac{\text{density of the material}}{\text{density of water}}$$

For example, if the density of copper is 8900 kg/m³,

$$\text{relative density of copper} = \frac{8900 \text{ [kg/m}^3]}{1000 \text{ [kg/m}^3]} = 8 \cdot 9.$$

* Derived units are obtained from the multiplication or division of base units, e.g. kg/m³ for density.

Since the relative density is a ratio of two quantities expressed in the same units, it is purely a number and therefore has no units.

Example 1.1 *A block of steel,* 200 *mm* × 100 *mm* × 60 *mm, has a mass of* 9·42 *kg. Calculate (a) the density and (b) the relative density of the steel.*

(a)   Volume of block $= 0·2 \text{ [m]} \times 0·1 \text{ [m]} \times 0·06 \text{ [m]}$
$$= 0·0012 \text{ m}^3,$$

∴   density of steel $= \dfrac{9·42 \text{ [kg]}}{0·0012 \text{ [m}^3]} = 7850 \text{ kg/m}^3.$

(b)   Since the density of water $= 1000 \text{ kg/m}^3$

∴   relative density of steel $= \dfrac{7850 \text{ [kg/m}^3]}{1000 \text{ [kg/m}^3]} = 7·85.$

1.5 Determination of density and relative density

(a) *Solids.* If the solid is in the form of a rectangular or cylindrical block, its dimensions can be measured and its volume calculated. The mass of the block can be determined by means of a beam balance such as an ordinary chemical balance, and the values of the density and of the relative density can then be calculated as in Example 1.1.

In the case of a solid having an irregular shape, a simple, though not very accurate, method of determining its *volume* is to fill vessel

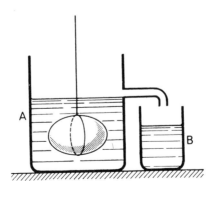

Fig. 1.1 Determination of the density of a solid.

A, shown in fig. 1.1, with water until it overflows. The object is then immersed in the water and the amount of water displaced is collected in vessel B. The mass of B is determined before and after the water is collected, and the difference in the readings gives the mass of water displaced. The volume of the solid can then be calculated from the fact that the volume of 1 kg of water is $0\cdot001$ m^3.

(*b*) *Liquids.* If a vessel of known volume is available, the mass of that volume of liquid is the difference between the mass of the vessel (i) empty and (ii) filled with the liquid. Hence the density of the liquid can be calculated.

The relative density of a liquid can be accurately determined by means of a special glass bottle known as a 'relative density bottle'. The glass stopper has a bore of small diameter centrally through it and fits accurately into the opening, as shown in fig. 1.2. Consequently, when the bottle is filled with a liquid and the stopper inserted, the surplus liquid flows out and the bottle will then contain a definite volume of the liquid.

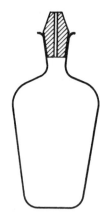

Fig. 1.2 A relative density bottle.

The procedure is to weigh the bottle after it has been dried internally and externally. It is then filled with water, the external surface is dried and the bottle is again weighed. The test is repeated with the liquid whose relative density is required.

Then, if x = mass of bottle empty,

y = mass of bottle filled with water

and z = mass of bottle filled with the

other liquid,

mass of water = $y - x$

and   mass of other liquid = $z - x$

$\therefore$ relative density of the $\left.\right\}$ = $\dfrac{z - x}{y - x}$.

other liquid

1.6 Force

Mechanics is largely concerned with force and the effects of force. Everyone who has exerted muscular force in lifting an object, pushing a load or overcoming any kind of resistance, has some conception of force.

Force is generally recognized and measured by its effects. Force exerted on a body tends to change its motion, and unless resisted by another force, will do so; for example, the pushing at a door by a hand, the pulling of a train by a locomotive and the acceleration of a falling body by the force of gravity.

Another effect which can occur when forces are exerted on a body is that the material of the body is deformed, i.e. its shape is altered. For example, it may be stretched, compressed, bent or twisted. Frequently, this deformation is so small as to be imperceptible to the unaided eye.

1.7 Unit of force

At this stage we can only deal with the subject in a brief and simple manner—full discussion will be found in Chapter 7. All we need say here is that the SI unit of force is termed the *newton* (symbol, N), to commemorate the name of Sir Isaac Newton, and is defined as *the force which, when applied to a mass of* 1 *kilogram gives it an acceleration of* 1 *metre per second every second* (*or* 1 m/s^2).

1.8 Mass and weight

The *mass* of a body is usually defined as the quantity of matter in the body, but the nature of mass cannot be fully appreciated without reference to its quality of *inertia* or reluctance to change of velocity. For instance, a cricket ball, thrown with a given force in a given direction, will travel a certain distance; but if a ball of the same size, made of lead, was thrown in the same direction with the same force, its acceleration and maximum velocity would be much less than those of the cricket ball, and it would not travel as far. This is due to the lead ball having a greater inertia or reluctance to acceleration than the cricket ball, and so we say that the mass of the lead ball is greater than that of the cricket ball.

A great English scientist, Sir Isaac Newton (1642-1727), showed that the earth attracts every body to itself with a force proportional to the mass of the body, and that the value of this force decreases as the body moves further away from the surface of the earth. The force with which a body is attracted towards the earth is termed the *weight* of that body, and its direction is always towards the centre of the earth.

Since the earth is not a perfect sphere, its radius being less at the poles than at the equator, the pull of the earth on a given mass is slightly different at different places on the earth's surface. For instance, if the scale of a *spring* balance is calibrated in London to read exactly 1 kilogram when a mass of 1 kg is suspended from the balance, then, if the balance and the same mass of 1 kg were taken to the north or south pole, the scale reading would be about 1·002 kg, whereas at the equator the reading would be about 0·997 kg.

Had the mass of 1 kg been placed in one pan of a beam balance, such as an ordinary chemical balance, and pieces of metal placed on the other pan until the beam was exactly balanced, then this equilibrium would hold whether the balance were used in London, at the equator, on the moon or at any point where the masses were subject to a gravitational force. This is due to the gravitational force at a given place being the same for the two pans and their contents. Hence it will be seen that the beam balance measures the *mass* of a body in comparison with the known mass of another body, and the result is independent of the value of the gravitational force at the place where the measurement is being made. A spring balance, on the other hand, measures the *weight* of the body; and its reading, for a given suspended mass, varies from place to place on the earth's surface and would be considerably less on the moon's surface.

When a mass of 1 kg rests on, say, a table as shown in fig. 1.3, it exerts a downward force on that table, but the magnitude of that force is different at different points on the earth's surface, as has already been mentioned. This means that if a body is allowed to fall freely (i.e. if the air resistance is negligible), the acceleration varies at different places. For instance, at sea level in

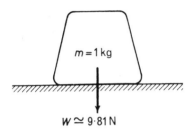

Fig. 1.3 Mass and weight.

the vicinity of London, gravitational acceleration is almost exactly 9.81 m/s^2, whereas at the equator the acceleration is about 9.780 m/s^2 and at each pole it is about 9.832 m/s^2. From the definition of the newton, it follows that the gravitational force at sea level on a mass of 1 kg (i.e. the weight of 1 kg) is almost exactly 9.81 newtons in London, 9.78 newtons at the equator and 9.832 newtons at each pole, as shown in fig. 1.4.

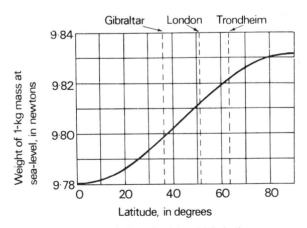

Fig. 1.4 Variation of weight with latitude.

It will be seen from fig. 1.4 that the weight of a 1-kg mass at sea level is within the range 9·81 N ± 0·1 per cent for latitudes extending from 38° to 61°, i.e. from the north coast of Africa to the central region of Norway, in the northern hemisphere, and for the corresponding region in the southern hemisphere. It follows that within this range of latitude, we can say that for most practical purposes:

$$\text{weight of a body} \simeq 9\cdot81\,m \text{ newtons,}$$

where m is the mass of the body in kilograms. The sign $\simeq$ will be used in this relationship, wherever it appears in this book, to indicate that '9·81' may be an approximate value and that the precise value depends upon the latitude at which the figure is used. Thus the force exerted on the table by the mass of 1 kg in fig. 1.3 is practically 9·81 N.

For very rough estimates, we could assume the weight of a body having a mass of m kilograms to be $10\,m$ newtons.

1.9 Pressure

The effect of a force applied to a surface depends on the area over which the force is applied. That is why nails and pins have sharp points, so the force is applied over a very small area. The term pressure is used for the force applied per unit area.

$$\text{pressure } p = \frac{\text{Force [newtons]}}{\text{area [metres}^2]}$$

The SI unit of *pressure* is the N/m^2. This is given a special name—the *pascal* (Pa).

Example 1.2. *A steel block having a mass of 80 kg rests on a table, the area of contact with the table being 2000 mm². Calculate* (a) *the downward force on the table and* (b) *the average pressure on the table in kilonewtons per square metre.*

(a)   Downward force

on table = weight of block

$$\simeq 80 \text{ [kg]} \times 9\cdot81 \text{ [N/kg]*} = 784\cdot8 \text{ N.}$$

---

* This expression can alternatively be stated thus:

weight of block $\simeq 80$ [kg] $\times 9\cdot81$ [m/s²] $= 784\cdot8$ N.

When the body under consideration is *stationary*, [N/kg] is better since it emphasizes the relationship between mass and weight, namely that a mass of 1 kg has a weight of approximately 9·81 N.

(b)   Area of contact $= 2000 \text{ mm}^2$

$$= \frac{2000 \text{ [mm}^2]}{1\,000\,000 \text{ [mm}^2/\text{m}^2]} = 0\cdot002 \text{ m}^2.$$

∴   average pressure
on table $=$ force per unit area

$$= \frac{\text{downward force}}{\text{area of contact}}$$

$$= \frac{784\cdot8 \text{ [N]}}{0\cdot002 \text{ [m}^2]} = 392\,400 \text{ Pa}$$

$$= 392\cdot4 \text{ kPa.}$$

1.10 Work

When a force is exerted against some form of resistance through a distance in the direction of the force, *work* is said to be done. For example, work is done if a body is lifted vertically upward against the gravitational pull of the earth; or if a spring is stretched, compressed or bent against the elastic resistance to deformation; or if a body is accelerated by a force applied to it.

The SI unit of work is the *joule** (symbol, J) to commemorate the English physicist James P. Joule (1818–89), famous for his experiments on the relationship between mechanical and thermal energies. The *joule* is defined as *the work done when a force of 1 newton is exerted through a distance of 1 metre in the direction of the force.* Hence, if a force F, in newtons, is exerted through a distance s, in metres, in the direction of the force,

work done, in joules $= F$ [newtons] $\times s$ [metres]

$$= Fs \qquad\qquad (1.1)$$

1 kilojoule [kJ] $= 1000$ J,
1 megajoule [MJ] $= 1\,000\,000$ J.

Example 1.3. *The work done in moving a body through a distance of 30 m is 600 J. Assuming the force to act in the direction of motion, calculate the average value of the force.*

From the expression (1.1), we have:

$$600 \text{ [J]} = F \times 30 \text{ [m]}$$

∴ $\qquad\qquad\qquad F = 20 \text{ N.}$

* *Joule* is pronounced 'jool', rhyming with 'tool'.

Example 1.4 *A body having a mass of 20 kg is lifted through a distance of 15 m. Calculate the work done.*

$$\text{Weight of the body} \simeq 20 \ [\text{kg}] \times 9\cdot81 \ [\text{N/kg}] = 196\cdot2 \ \text{N},$$
$$\therefore \qquad \text{work done} = 196\cdot2 \ [\text{N}] \times 15 \ [\text{m}] = 2943 \ \text{J}$$
$$= 2\cdot943 \ \text{kJ}.$$

1.11 Power

Power is defined as *the rate of doing work* and is a measure of the work done per second. The SI unit of *power* is the *watt* (symbol, W), named after the famous Scottish engineer James Watt (1736-1819). *The watt is equal to 1 joule per second.*

$$1 \text{ kilowatt } [\text{kW}] = 1000 \text{ W},$$
$$1 \text{ megawatt } [\text{MW}] = 1\,000\,000 \text{ W}.$$

Example 1.5 *A horizontal force of 60 N is applied to a body to move it at a uniform velocity through a distance of 20 m in 8 s in the direction of the force. Calculate the value of the power.*

$$\text{Work done} = 60 \ [\text{N}] \times 20 \ [\text{m}] = 1200 \ \text{J},$$
$$\therefore \qquad \text{power} = \text{work done per second}$$
$$= \frac{1200 \ [\text{J}]}{8 \ [\text{s}]} = 150 \text{ W}.$$

Example 1.6 *Calculate the power required to lift a mass of 300 kg at a constant speed through a vertical height of 200 m in 4 min.*

$$\text{Force required to lift load} \simeq 300 \ [\text{kg}] \times 9\cdot81 \ [\text{N/kg}] = 2943 \ \text{N}.$$
$$\text{Work done in 4 min} = 2943 \ [\text{N}] \times 200 \ [\text{m}]$$
$$= 588\,600 \ \text{J}.$$
$$\therefore \qquad \text{power} = \frac{588\,600 \ [\text{J}]}{(4 \times 60) \ [\text{s}]} = 2453 \text{ W}$$
$$= 2\cdot453 \text{ kW}.$$

Summary of Chapter 1

The SI units of length, mass and time are the metre, the kilogram and the second respectively.

Density is the mass per unit volume and the relative density of a material is the ratio of the density of that material to the density of water.

The SI unit of force is the newton, namely the force required to give a mass of 1 kg an acceleration of 1 m/s².

Pressure is force per unit area and the SI unit of pressure is the pascal (Pa), where 1 Pa = 1 N/m².

The SI unit of work is the joule. It is the work done when a force of 1 N is exerted through a distance of 1 m in the direction of the force.

Work done, in joules = $F s$ (1.1)

Power is the rate of doing work and the SI unit of power is the watt which is 1 joule/second.

EXAMPLES 1

1. Calculate the area, in square metres, of: (*a*) a square of side 600 mm, (*b*) a triangle of base 400 mm and height 500 mm and (*c*) a circle of diameter 1·5 m.
2. Calculate the volume, in cubic metres, of: (*a*) a room, 7 m × 4 m × 3 m and (*b*) a sphere having a diameter of 1·5 m.
3. Calculate the length of each side of a cube having a volume of 3 m³.
4. Calculate the diameter of a sphere having a volume of 0·8 m³.
5. A steel plate measures 500 mm × 200 mm × 10 mm. If the relative density of the steel is 7·8, calculate the mass of the plate.
6. If the density of aluminium is 2700 kg/m³, what is the volume, in cubic centimetres, of 5 kg of this material?
7. A block of wood, measuring 300 mm × 150 mm × 60 mm, has a mass of 2·3 kg. Calculate the density of the wood.
8. If the mass of 7500 mm³ of lead is 85 g, calculate the relative density of the lead.
9. A 2-m length of iron pipe has a mass of 137 kg. The external and internal diameters of the pipe are 160 mm and 120 mm respectively. Calculate (*a*) the density and (*b*) the relative density of the iron.
10. Hydrogen has a density of 89·9 g/m³ at a certain temperature and a certain pressure. Calculate the volume, in cubic metres, of hydrogen having a mass of 0·1 kg at the same temperature and pressure.
11. A 10-kg mass is suspended at the end of a cord. Calculate the approximate value of the force, in newtons, exerted by the cord.
12. A body is suspended from a spring balance. The reading on the balance is 4 N. Assuming the local gravitational force to be 9·81 N/kg, calculate the mass of the body in grams.
13. A vertical column supports a body having a mass of 6 Mg (or 6 tonnes).

Calculate the approximate value of the upward force, in kilonewtons, exerted by the column.

14. A cylindrical vessel has an internal diameter of 100 mm and an internal length of 300 mm. It is stood on end and filled with oil having a relative density of 0·8. Calculate (*a*) the total thrust on the bottom of the cylinder and (*b*) the pressure on the bottom of the cylinder, in kilonewtons per square metre.

15. A force of 30 N acts through a distance of 4 m in the direction of the force. What is the work done?

16. A force of 700 N does 2500 J of work. Through what distance does the force move, assuming that it acts in the direction of motion?

17. Calculate the force, in kilonewtons, required to do 3 MJ of work when it moves through a distance of 200 m in the direction of the force.

18. A horizontal force of 20 N is applied to a body to move it at a uniform velocity through a distance of 80 m in 6 s in the direction of the force. Calculate (*a*) the work done and (*b*) the power.

19. The work done by a force in moving a body at a uniform speed through a distance of 20 m, in the direction of the force, is 3 kJ. If this work is done in 10 s, calculate (*a*) the force and (*b*) the power.

20. The power required to lift a certain body through a distance of 30 m in 6 s is 200 W. Calculate (*a*) the work done and (*b*) the mass of the body.

21. If 40 MJ of work are done in lifting 5 Mg (or 5 t) of coal, through what height is it lifted? If the time taken to lift the 5 Mg of coal is 120 s, what is the power required? Neglect any losses.

22. If 60 m³ of water are pumped per hour to a height of 80 m, calculate (*a*) the power required, in kilowatts, and (*b*) the work done, in megajoules, in 10 min. Neglect any losses and assume that 1 m³ of water has a mass of 1000 kg.

ANSWERS TO EXAMPLES 1

1. 0·36 m², 0·1 m², 1·765 m².
2. 84 m³, 1·77 m³.
3. 1·44 m.
4. 1·15 m.
5. 7·8 kg.
6. 1850 cm³.
7. 852 kg/m³.
8. 11·3.
9. 7800 kg/m³, 7·8.
10. 1·112 m³.
11. 98·1 N.

12. 408 g.
13. 58·86 kN.
14. 18·5 N, 2·354 kPa.
15. 120 J.
16. 3·57 m.
17. 15 kN.
18. 1600 J, 267 W.
19. 150 N, 300 W.
20. 1200 J, 4·08 kg.
21. 815 m, 333 kW.
22. 13·08 kW, 7·848 MJ.

CHAPTER 2

Triangle and polygon of forces

2.1 Forces and reactions

One of the most important facts to be appreciated in mechanics is that to every force acting on one body there is an equal and opposite force or reaction acting on another body. This is easily stated but not so readily fully understood, but consideration of simple examples will help.

Suppose a piece of metal to be suspended by a string as shown in fig. 2.1. The *weight* of the metal exerts a *downward* pull on the string, and the string exerts an *upward* force on the piece of metal.

Fig. 2.1 Force exerted on a body.   Fig. 2.2 Force exerted by a body

Since the body is stationary, it is in a state of *equilibrium*, and force F is therefore exactly equal in magnitude and opposite in direction to force W. A body cannot remain in equilibrium under the action of a *single* force. For instance, if the string in fig. 2.1 were cut, the only force acting on the metal would be its weight, and the metal would fall to the ground, its speed increasing as it fell.

The forces shown in fig. 2.1 are those exerted *on* the piece of metal. The force of gravity W is acting downward; and force F is therefore acting upward and is the *reaction* to the downward force W exerted *by* the piece of metal on the string, as shown in fig. 2.2.

Let us next consider a rigid* bar AB, fig. 2.3(a). Suppose the bar to be pulled by two forces F applied, say, by helical springs attached to the loops at its ends. Force F is transmitted by the material throughout the whole length of the bar. If we consider any short length CD of the bar, it is pulled at C by a force F, acting towards A, exerted by the material in the bar lying to the left of C. It is also pulled at D by an equal force F, acting towards B, exerted by the material to the right of D. Also, piece CD, when forming part of the whole bar, as in fig. 2.3(a), exerts a rightward pull F at C on the portion AC of the bar, and a leftward pull F at D on portion DB of the bar, as shown in fig. 2.3(b).

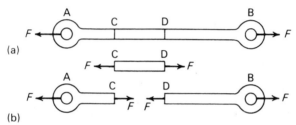

Fig. 2.3 Tension in a bar.

It will be seen that at any cross-section of the bar there are equal and opposite forces F. Each of the two pieces, into which any cross-section divides the bar, pulls the piece on the opposite side of the section towards itself. It is this dual action which constitutes *tension*.

The forces F shown in fig. 2.3(a) are those exerted outwards *on* the bar *by* the helical springs at A and B. The forces exerted *by* the bar *on* the springs are equal to F but are inwards. When arrowheads are used to indicate forces in such a bar, it is necessary to state whether we are indicating the forces exerted *on* the bar or those exerted *by* the bar on the springs or other means used for exerting the tension.

2.2 Representation of a force by a vector
A force has the following characteristics:
(a) magnitude,
(b) direction (line of action and sense),
(c) point of application.

* A rigid body is a body whose size and shape are unaffected by forces acting upon it. No solid is perfectly rigid, but we assume it is so unless otherwise stated.

For example, in fig. 2.1, the line of action of the force F exerted by the string on the suspended metal is vertical and its sense is upwards. The magnitude of this force may be, say, 30 N and is the same at all points along the length of the string.

A quantity which has magnitude and direction is referred to as a *vector quantity*, whereas a quantity which has magnitude only is known as a *scalar quantity*. Quantities such as mass and time possess magnitude only and therefore are scalar quantities. Force, on the other hand, possesses both magnitude and direction and is consequently a vector quantity.

In the case of a body suspended from a string (fig. 2.1), it is not sufficient to state that the line of action of force F is vertical; we must also state or indicate that this force is acting upwards, i.e. the *sense* of the force must be specified.

A vector quantity can be represented by a straight line whose length represents to a known scale the magnitude of the quantity, and whose direction is parallel to the line of action of that quantity. Such a line is termed a *vector*. For example, if the *weight*, W, of the suspended mass in fig. 2.1 is 30 N, then, using a scale of, say 1 mm to 1 N, we can represent this weight by a vertical line ab, 30 mm long, as shown in fig. 2.4(a). The direction of this force is indicated by the arrowhead. Also, when we say that the force is represented by vector ab, we mean that the direction of the force is downward from a to b, not upward from b to a.

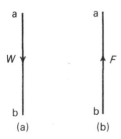

Fig. 2.4 Representation of a force by a vector.

The force F exerted *on* the mass *by* the string is equal and exactly opposite to force W and is therefore represented to the same scale by vector ba in fig. 2.4(b); i.e. vector ba represents a force of 30 N acting vertically upward.

2.3 Equilibrium

When two or more forces act upon a body and are so arranged that the body remains at rest or moves at a constant speed in a straight line, the forces are said to be in *equilibrium*.

It was mentioned in section 2.1 that a body cannot remain in equilibrium under the action of a single force. The effect of a single force on a body is to accelerate the body in the direction of the force.

The simplest case of equilibrium is that of two equal and opposite forces in the same straight line, such as the weight W of the piece of metal acting vertically downward in fig. 2.1, and the equal upward reaction or supporting force F exerted by the string. In this case, the body remains at rest.

The forces may be two equal and opposite *pulls*, as in figs. 2.1 and 2.3, in which case the body is said to be in *tension*. On the other hand, the forces may be two equal and opposite *pushes*, in which case the body is said to be in *compression*. For instance, if a helical spring, anchored at one end, has a push applied at the other end, the spring is compressed; and the push, together with the equal and opposite reaction exerted by the anchorage at the other end, are compressive forces.

Another example of the equilibrium of two equal and opposite forces is that of a locomotive pulling a train at a *constant* speed along a straight level track. The forward pull exerted by the locomotive on the train is equal to the backward drag exerted by the train on the locomotive. This backward drag is due to friction at the axles and the rails together with some air resistance. If the engine were to exert a larger pull than the drag due to friction and air resistance, the forces would not be in equilibrium and the train would accelerate.

In this chapter, we shall confine our attention to the equilibrium of forces acting on a body which remains stationary.

We shall next consider the condition for the equilibrium of a body subjected to three inclined co-planar forces, i.e. three non-parallel forces acting in the same plane.

2.4 Three inclined forces: Triangle of Forces

Fig. 2.5 shows three strings in the same horizontal plane, knotted together or attached to a small ring. Spring balances P, Q, and R, calibrated in newtons, are attached to the respective strings.

Suppose the strings attached to balances P, Q and R to be subjected to tensions of 5 N, 4 N and 6 N respectively. The directions of the three strings are marked on a sheet of paper pinned to a board behind the strings. The paper is then removed and a triangle *cab* is constructed by drawing side *ca* parallel to the string attached to balance R, and making *ca* 120 mm long to represent the force of 6 N at a scale of 1 mm to 0·05 N. The triangle is then completed by drawing lines from *a* and *c* respectively parallel to the directions of the other two strings to meet at *b*. On measuring these two sides, it is found that *ab* represents 5 N and *bc* represents 4 N to the same scale as that used for *ca*. The degree of accuracy in this experiment will depend upon the care exercised and the quality and condition of the spring balances.

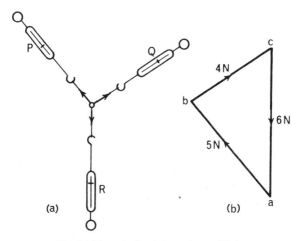

Fig. 2.5 Three inclined forces in equilibrium.

The relationship demonstrated* by this experiment is known as the Triangle of Forces Theorem which can be expressed thus:

If three forces, acting at a point, can be represented in magnitude and direction by the sides of a triangle taken in a cyclic order around the triangle, the forces are in equilibrium.

Conversely, *if three forces, acting at a point, are in equilibrium, they can be represented in magnitude and direction by the sides of a triangle taken in a cyclic order.*

* The Triangle of Forces Theorem can be *proved* by the aid of the Principle of Moments (section 3.4).

B

It is this converse theorem which is very useful for solving problems on the equilibrium of three non-parallel forces acting at a point.

2.5 Bow's notation

Before we proceed to the application of the triangular force diagram to problems, it is necessary to state the way in which forces are named by letters.

Suppose three inclined forces, *P, Q* and *R*, acting at a point O, to be in equilibrium, and suppose the lines of action of these three forces to be as shown in fig. 2.6(a). Such a diagram is referred to as a *space* diagram. The forces in this space diagram can be designated by capital letters inserted in a clockwise direction in the *spaces* between the forces, as indicated in fig. 2.6 (a). Force *P* can then be referred to as force AB, force *Q* as force BC and force *R* as force CA.

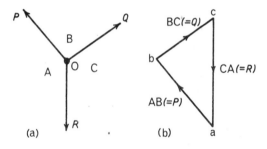

Fig. 2.6 Bow's notation.

The force or vector diagram of fig. 2.6(b) is constructed by drawing vector *ab* parallel to the line of action of force AB to represent to some convenient scale the magnitude of force AB, having its direction *a* to *b* (not *b* to *a*).

The remainder of the vector triangle of fig. 2.6(b) is constructed by drawing two lines, one from *b* parallel to force BC and the other from *a* parallel to force CA, meeting at point *c*.

From a comparison of figs. 2.6(a) and (b), it will be seen that forces AB, BC and CA in the space diagram are represented by vectors *ab*, *bc* and *ca* respectively in the force diagram. This system of lettering is known as Bow's notation.

2.6 Resultant

Consider for example, in fig. 2.7, a body suspended by two cords from D and E. Suppose the gravitational force on this body (i.e. the weight W of the body) to be 80 N. The tension in the cord supporting the body is therefore 80 N and the arrowhead on this cord indicates the direction of the force exerted *by* the cord on junction O. Suppose the

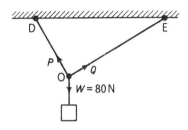

Fig. 2.7 Equilibrant of two inclined forces.

corresponding tensions in cords OD and OE to be P and Q respectively. The arrowheads inserted on OD and OE in fig. 2.7 indicate the directions of the forces exerted *by* the cords on junction O.

Since the two forces P and Q balance the vertical downward pull of 80 N, they may be said to exert on junction O a vertical *upward* pull of 80 N which just balances the downward equilibrant of P and Q, being equal and opposite to it and in the same straight line. Such a force is termed the *resultant* of forces P and Q; for the purpose of calculating the effects of two forces, it is often convenient to replace the two forces by their resultant. Thus we can replace P and Q of fig. 2.7 by an upward resultant force R of 80 N, as shown in fig. 2.8. This resultant exactly balances the downward force of 80 N, namely the weight of the body.

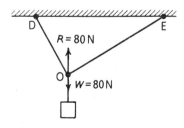

Fig. 2.8 Resultant.

Fig. 2.9(a) represents the space diagram for the three forces of fig. 2.7, and fig. 2.9(b) represents the force or vector diagram for the *three* forces. The term space diagram is used for a pictorial representation of the problem.

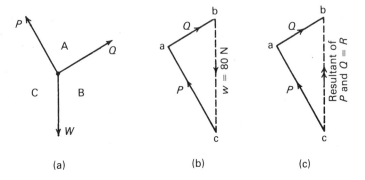

Fig. 2.9 Equilibrant and resultant.

The dotted line *bc* in fig. 2.9(b) represents the force that has to be applied to cancel out the effects of forces *P* and *Q* and is called the equilibrant of *P* and *Q*.

The dotted line *bc* in fig. 2.9(c) has two arrowheads in the direction *c* to *b* to indicate that this vector represents the *resultant* *R* of the *two* forces *P* and *Q*. In other words, vector *cb* represents a *single* force *R* that will replace and produce the same effect as the combination of forces *P* and *Q*.

2.7 Parallelogram of forces

The magnitude and direction of the resultant of the two forces *P* and *Q* in fig. 2.7 can also be determined by the construction shown in fig. 2.10, where vectors OA and OB are drawn to scale to represent forces *P* and *Q* in magnitude and direction.

The parallelogram OACB is completed by drawing AC parallel to OB and BC parallel to OA. The diagonal OC of this *parallelogram of forces* represents in magnitude and direction the resultant of the two forces *P* and *Q*.

Two arrowheads have been inserted on OC in fig. 2.10 to indicate that this vector is *not a third force* but is a *single force* that can replace the *two* forces *P* and *Q*.

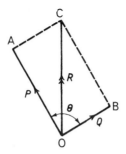

Fig. 2.10 Parallelogram of forces.

If θ be the angle between the directions of the two forces P and Q, the resultant force is given by:

$$R^2 = P^2 + Q^2 + 2PQ \cos \theta$$

or $$R = \sqrt{(P^2 + Q^2 + 2PQ \cos \theta)}$$

Comparison of figs. 2.9(c) and 2.10 shows that triangle *abc* of fig. 2.9(c) is exactly similar to triangle ACO of fig. 2.10. They differ only in one respect: the three lines OA, OB and OC in fig. 2.10 represent the relative lines of action, whereas line *ab* in fig. 2.9(c) shows force Q displaced relative to its true line of action.

The choice of the triangle or the parallelogram for use in determining the resultant of two forces is largely a matter of personal preference.

Example 2.1 *A mass of 10 kg is suspended by two cords from points D and E, as shown in fig. 2.11(a). Calculate the tension in each cord.*

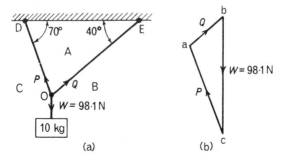

Fig. 2.11 Example 2.1.

Weight of the 10-kg mass $= W \simeq 10 \times 9{\cdot}81$
$$= 98{\cdot}1 \text{ N.}$$

Draw a vertical line bc in fig. 2.11(b) to represent W to a scale of, say, 1 mm to 1 N. Draw a line from b parallel to force Q; and draw a line from c parallel to force P to meet the first line at a.

By measurement, $ca = 80$ mm and $ab = 35{\cdot}7$ mm.

Hence,   tension in cord OD $= P = 80$ N

and   tension in cord OE $= Q = 35{\cdot}7$ N.

Example 2.2 *A jib crane has a jib 5 m long and a tie-rod 3·5 m long attached to a post 2 m vertically above the foot of the jib, as shown in fig. 2.12. Determine the forces transmitted by the jib and the tie-rod when a mass of 3 tonnes is suspended from the crane head.*

The problem is set out in fig. 2.12(a), and fig. 2.12(b) represents the force or vector triangle *abc* for the equilibrium of the joint at the crane head for the three forces AB, BC and CA acting at that point.

Mass suspended at crane head $= 3$ t $= 3000$ kg,

$\therefore$ force BC $=$ weight of suspended mass

$$\simeq 3000 \times 9{\cdot}81 = 29\ 430 \text{ N}$$

$$= 29{\cdot}43 \text{ kN.}$$

This force BC is set off to a scale of, say, 1 mm to 0·5 kN downward at bc in fig. 2.12(b), i.e. line bc is drawn 58·8 mm long. The directions of ca and ab are known and it is only necessary to draw lines parallel to the jib and the tie-rod from c and b respectively to meet at a to complete the force triangle *abc*. By measurement, ca is found to represent 73·6 kN and ab to represent 51·5 kN. Also, since the force of 29·43 kN (BC) acts downwards, i.e. from b to c, the sequence of letters in the force triangle is anticlockwise. Hence the force in the jib (at the crane head) acts in the direction c to a and *thrusts* at the head. Consequently the jib is in compression. The force in the tie-rod acts from a to b and therefore *pulls* at the head. Hence the tie-rod is in tension.

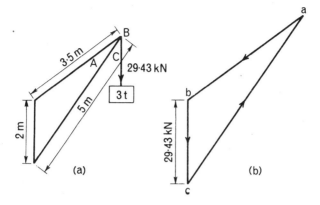

Fig. 2.12 Forces at head of crane.

It will be noticed that the triangular force diagram is geo-
metrically similar to the triangular skeleton diagram of the crane
in the space diagram. Hence, from a mere sketch of the force
diagram, not drawn to scale, we could easily calculate the forces.
Thus, $ca = 29.43$ [kN] $\times 5/2 = 73.6$ kN
and $ab = 29.43$ [kN] $\times 3.5/2 = 51.5$ kN.

Example 2.3 *A piece of machinery having a mass of 2 t is lifted by a
crane hook attached to the middle point of a rope sling as in fig. 2.13.
The hook is vertically above the centre of gravity* of the load and the
sling is symmetrically placed as shown. Determine the tension in the
two sides of the sling and examine the effect of shortening the sling.*

Mass of machinery = 2 t = 2000 kg,
∴   weight of machinery ≃ 2000 × 9·81 = 19 620 N
= 19·62 kN.

Insert the letters A, B and C in the spaces around the junction of
the forces at the hook in fig. 2.13(a) and draw the force triangle,
fig. 2.13(b), for the three forces which are in equilibrium at the
hook. The tension in the rope above the hook must obviously
be 19·62 kN.

---

* The centre of gravity is the point at which the entire weight of the body is considered to act, see Section 3.10.

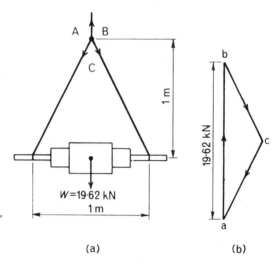

Fig. 2.13 Example 2.3

The force diagram is constructed by first drawing *ab* vertically to represent the force of 19·62 kN to a convenient scale and then drawing from *a* a line parallel to force AC and from *b* a line parallel to force BC. These two lines meet at *c*. The length *bc* represents to scale the pull BC on the hook. By measurement, it is found that *bc* represents 10·95 kN. By symmetry, the pull CA is also 10·95 kN.

If the rope sling were longer, the inclination of the sling to the horizontal would be steeper and *c* would lie nearer to *ab*. Consequently the tension in the sling would be reduced towards half the load. On the other hand, if the sling were shortened, the inclination would be smaller and *c* would lie further from *ab*, so that the tension in the rope would be greater. If the sling becomes nearly horizontal, the tension becomes very great (for *c* is then a long way to the right). Thus a sling or a picture cord may easily be broken by being too nearly at right angles to the direction of the load.

Example 2.4 *A triangular roof frame consists of a horizontal tie-rod and two rafters inclined at 30° and 60° respectively to the horizontal and is supported at the ends of its span as in fig. 2.14(a). If it carries a load of 4 kN at the apex, determine the resulting thrusts in the rafters, the tension in the tie-rod and the reactions at the supports.*

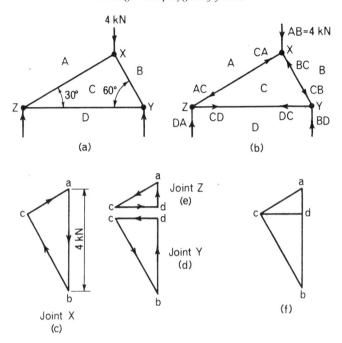

Fig. 2.14 Example 2.4.

Letter the spaces, A, B, C and D, as shown in fig. 2.14(a). The downward force AB on joint X is 4 kN and *this is the only force whose value is known*. In the space diagram of fig. 2.14(b), the arrow-heads indicate the directions of the various forces at each of the three joints; thus, BC represents the thrust in rafter XY, acting towards joint X. At the other end of this rafter, there is a thrust CB of equal magnitude, acting towards joint Y. Similarly DC represents the pull exerted by the tie-rod YZ on joint Y and CD represents the pull on joint Z. It will be noted that the sequence of each pair of letters is clockwise relative to the respective joints.

Joint X. The force triangle of fig. 2.14(c) is constructed by drawing vector *ab* to represent the downward load AB of 4 kN to a convenient scale. Vectors *bc* and *ac* are then drawn parallel to rafters XY and XZ respectively. By measurement,

> thrust in rafter XY towards joint X $= bc = 3\cdot46$ kN

and   thrust in rafter XZ towards joint X $= ca = 2$ kN.

Joints Y and Z. By a procedure similar to that used for joint X, force triangles *cbd* and *acd* of figs. 2.14(d) and 2.14(e) are drawn to represent to scale the forces acting at joints Y and Z respectively. Vectors *cb* and *ac* in these diagrams have the same magnitude as vectors *bc* and *ca* respectively in fig. 2.14(c) and are drawn parallel to them. From triangle *cbd* it is found that for joint Y:

thrust in rafter XY towards joint Y = *cb* = 3·46 kN,

upward reaction of support at Y = *bd* = 3 kN

and   pull of tie-rod YZ on joint Y = *dc* = 1·73 kN.

Similarly, from triangle *acd* it is found that for joint Z:

thrust in rafter XZ towards joint Z = *ac* = 2 kN,

upward reaction of support at Z = *da* = 1 kN

and   pull of tie-rod YZ on joint Z = *cd* = 1·73 kN.

It is evident that the sum of the upward reactions at joints Y and Z, namely 3 kN + 1 kN = 4 kN, must be equal but opposite in direction to the vertical load of 4 kN at joint X.

The triangles of figs. 2.14(c), (d) and (e) can be combined in one diagram as shown in fig. 2.14(f); but it is not possible to insert arrowheads on these vectors since their directions are opposite for the three joints.

2.8 Equilibrium of three inclined forces: concurrency

If a body is in equilibrium under the action of three inclined forces in the same plane, the lines of action of the three forces must pass through one point, i.e. the forces are said to be concurrent.

The truth of the concurrency of three non-parallel forces acting in one plane can easily be demonstrated by attaching three strings, A, B and C, to a light sheet of metal or plywood D, as in fig. 2.15. A piece of metal is suspended from string C, and strings A and B can be pulled in various directions. For each test, lines representing the directions of strings A, B and C are drawn on the sheet, as shown dotted in fig. 2.15. In every case it is found that these lines meet at a point. Similarly it may be demonstrated that concurrency is *not* necessary for the equilibrium of four or more forces in equilibrium.

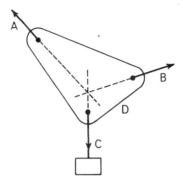

Fig. 2.15 Concurrency of three non-parallel forces in equilibrium.

As an example of the application of concurrency in the equilibrium of three inclined forces, let us consider a uniform ladder AB, fig. 2.16, resting against a *smooth* vertical wall with its lower end on rough ground.

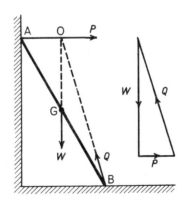

Fig. 2.16 Equilibrium of a ladder.

To determine the supporting forces exerted by the wall and the ground, we have to consider the following factors.

(*a*) Since the ladder is assumed uniform, its weight *W* can be regarded as acting vertically downwards through a point G midway between its ends.

(*b*) The vertical wall is assumed perfectly smooth, hence there are no frictional forces and consequently the reaction P exerted by the wall must be perpendicular to the wall. Hence the direction of P is horizontal. The line representing the direction of the reaction at A meets the vertical line through G at a point O.

(c) The third force acting on the ladder is the reaction Q of the ground at B. For the ladder to be in equilibrium, it is necessary for the three forces W, P and Q to be concurrent. Consequently the line of action of Q must pass through point O and the direction of Q is therefore given by line BO. This force Q is the resultant of two forces, the reaction and a frictional force, acting at B.

2.9 Resolution of a force into two components

In sections 2.6 and 2.7 we considered how two forces could be replaced by a single force equal to the vector sum of the two forces. Conversely, we can replace a single force by two forces such that their vector sum is equal to the original force in magnitude and direction. Such a process is termed *resolution* of a force, i.e. splitting up the force into components.

The three diagrams of fig. 2.17 show how the two components P and Q of a force R can be in any direction—the only condition being that the vector sum of P and Q must be the same as the vector representing R. In practice, the directions of the components are usually specified, and these directions are generally at right angles to each other, as shown in fig. 2.17(c). P and Q are then referred to as the *rectangular components* of R.

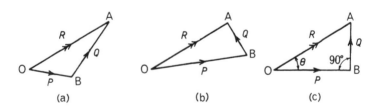

Fig. 2.17 Resolution of a force.

If θ be the angle between the original force R and the horizontal component P in fig. 2.17(c), then:

horizontal component $= P = R \cos \theta$

and vertical component $= Q = R \sin \theta$.

Example 2.5 *A barge is being towed along a canal by a rope inclined at 20° to the direction of the canal. The tension in the rope is 200 N. Calculate the values of the rectangular components of this tension, one component being in the direction of the canal.*

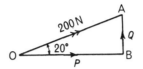

Fig. 2.18 Example 2.5.

In fig. 2.18, OA represents the tension of 200 N in the rope. Then component force P in the direction of the canal

$$= \text{OB} = \text{OA} \cos 20°$$
$$= 200 \times 0.94 = 188 \text{ N,}$$

and component Q perpendicular to the direction of the canal

$$= \text{BA} = \text{OA} \sin 20°$$
$$= 200 \times 0.342 = 68.4 \text{ N.}$$

Example 2.6 *A force of 60 N acts horizontally. Resolve this force into two components, one of which acts at an angle of 40° above the horizontal and the other at 20° below the horizontal. Determine the value of each component.*

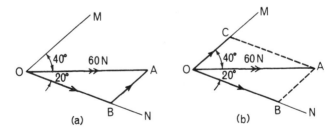

Fig. 2.19 Example 2.6.

Suppose vector OA in fig. 2.19(a) to represent the horizontal force of 60 N to a convenient scale. Lines OM and ON are drawn at 40° above and 20° below the horizontal respectively. From A, draw a line parallel to OM, cutting ON at B; then OB and BA

represent the required components. From the scaled diagram it is
found that component OB is 44·5 N and component BA is 23·7 N.

An alternative construction is to draw a line from A parallel to
ON, cutting OM at C, and another line parallel to OM, cutting ON
at B, as shown in fig. 2.19(b). The component forces are represented
in magnitude and direction by OB and OC. Their values are, of
course, the same as those already determined from fig. 2.19(a).

Example 2.7 *Four co-planar forces act at a point O, the values and
directions of the forces being as shown in fig. 2.20. Calculate the
value and direction of the resultant force.*

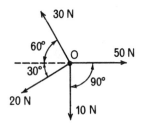

Fig. 2.20 Example 2.7.

It will be assumed that the horizontal components of the various
forces are positive when they are acting towards the right of point
O, and that their vertical components are positive when they are
acting upwards from O.

For the 50-N force,

$$\text{horizontal component} = 50 \cos 0° = 50 \text{ N}$$

and $$\text{vertical component} = 50 \sin 0° = 0.$$

For the 10-N force,

$$\text{horizontal component} = 10 \cos 90° = 0$$

and $$\text{vertical component} = -10 \sin 90° = -10 \text{ N}.$$

For the 20-N force,

$$\text{horizontal component} = -20 \cos 30° = -17·32 \text{ N}$$

and $$\text{vertical component} = -20 \sin 30° = -10 \text{ N}.$$

For the 30-N force,

$$\text{horizontal component} = -30 \cos 60° = -15 \text{ N}$$

and $$\text{vertical component} = 30 \sin 60° = 25\cdot98 \text{ N}.$$

Hence,

$$\text{resultant horizontal component} = 50 - 17\cdot32 - 15$$
$$= 17\cdot68 \text{ N}$$

and $$\text{resultant vertical component} = -10 - 10 + 25\cdot98$$
$$= 5\cdot98 \text{ N}.$$

The above calculation can be tabulated thus:

Force [N]	Horizontal component [N]	Vertical component [N]
50	$50 \cos 0° = 50\cdot0$	$50 \sin 0° = 0$
10	$10 \cos 90° = 0$	$-10 \sin 90° = -10\cdot0$
20	$-20 \cos 30° = -17\cdot32$	$-20 \sin 30° = -10\cdot0$
30	$-30 \cos 60° = -15\cdot0$	$30 \sin 60° = 25\cdot98$
Resultant	$= 17\cdot68$	$= 5\cdot98$

The resultant horizontal and vertical components are represented by OA and AB respectively in fig. 2.21, and the resultant force R is represented by OB.

Since $$OB^2 = OA^2 + AB^2$$
∴ $$R^2 = (17\cdot68)^2 + (5\cdot98)^2 = 348\cdot4$$
so that $$R = 18\cdot67 \text{ N}.$$

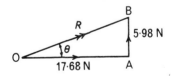

Fig. 2.21 Horizontal and vertical components.

If θ be the angle between the resultant and the horizontal component,

$$\tan \theta = \frac{AB}{OA} = \frac{5\cdot98}{17\cdot68} = 0\cdot338,$$
$$\therefore \theta = 18\cdot7°.$$

2.10 Polygon of forces

The Polygon of Forces Theorem is an extension of the Triangle of Forces Theorem and can be expressed thus:

If four or more co-planar forces, acting at a point, can be represented in magnitude and direction by the sides of a polygon taken in a cyclic order around the polygon, the forces will be in equilibrium.

Conversely, *if four or more co-planar forces acting at a point are in equilibrium, the magnitudes and directions of these forces can be represented by the sides of a polygon taken in a cyclic order around the polygon.*

Fig. 2.22(a) is a space diagram showing the directions of five forces, *P*, *Q*, *R*, *S* and *T*, in one plane, acting at a point. Using Bow's notation, we can draw the force diagram shown in fig. 2.22(b), where vector *ab* represents the value and direction of force *P*. From *b*, draw vector *bc* to represent the value and direction of *Q*. Then the dotted line *ac* represents the resultant of *P* and *Q*, as already explained in section 2.6.

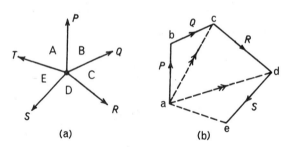

Fig. 2.22 Polygon of forces.

Similarly, from *c*, draw vector *cd* to represent force *R*; then the dotted line *ad* represents the resultant of *P*, *Q* and *R*. Finally, from *d*, draw vector *de* to represent force *S*. Vector *ae* is the resultant of vectors *ad* and *de* and is therefore the resultant of forces *P*, *Q*, *R* and *S*.

If this vector *ae* is *equal in magnitude* but *opposite in direction* to force *T* in fig. 2.22(a), then *T* is the equilibrant of forces *P*, *Q*, *R* and *S*, and the *five* forces are in equilibrium. In other words, if the five forces shown in the space diagram of fig. 2.22(a) are represented by vectors forming a *closed* polygon as in fig. 2.22(b), these forces are in equilibrium.

Example 2.8 *Four forces, in one plane, act at a point O, the values and directions of the forces being as shown in fig. 2.23(a). Determine by the polygon of forces the value and direction of the resultant force.*

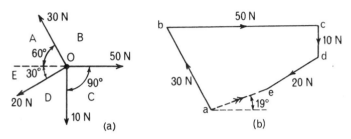

Fig. 2.23 Example 2.8.

The data are exactly the same as those of Example 2.7.

Fig. 2.23(a) is the space diagram of the four forces. Since these forces are not meant to be in equilibrium, it is necessary to add an additional capital letter E in the same space as that occupied by letter A, but A is placed near the 30-N force and E near the 20-N force.

Using a scale of, say 1 mm to 1 N, draw the force diagram of fig. 2.23(b), starting from *a*. The resultant of the four forces is represented by vector *ae*. By measurement, the value of this resultant is 18·6 N and its direction is 19° north-east of the 50-N force. These figures are in agreement with those calculated in Example 2.7.

The equilibrant of the four forces is represented by *ea* and its magnitude 19° south-west of the east-west line passing through point O. If an *additional* force of this magnitude were applied in this direction at O, the *five* forces would be in equilibrium.

Summary of Chapter 2

A body cannot remain in equilibrium under the action of a *single* force.

A body is in equilibrium under the action of *two* forces provided that the forces are equal in magnitude and exactly opposite in direction.

Triangle of Forces Theorem. If *three* inclined forces, acting at a point, can be represented in magnitude and direction by the sides of a triangle taken in a cyclic order, the forces will be in equilibrium; or conversely, if the three inclined forces are in equilibrium, their magnitudes and directions can be represented by the sides of a triangle, taken in a cyclic order around the triangle.

If three inclined forces are in equilibrium, they must be concurrent, i.e. their lines of action must pass through the same point.

A force can be replaced by two components of which it would be the resultant. Usually the two components are at right angles to each other.

Polygon of Forces Theorem. If *four* or more co-planar forces, acting at a point, can be represented in magnitude and direction by the sides of a polygon taken in cyclic order, the forces will be in equilibrium; or conversely, if four or more co-planar forces are in equilibrium, their magnitudes and directions can be represented by the sides of a polygon, taken in a cyclic order around the polygon.

EXAMPLES 2

1. Three forces, A, B and C, act in the same vertical plane from a point O. Force A is 30 N and acts horizontally to the left of O. Force B is 40 N and acts vertically upwards. Determine with the aid of a force diagram the value of force C and the direction in which it acts, if the forces are in equilibrium.

2. Three cords, A, B and C, are attached to one another at a point O. The angle between cords A and B is 80° and that between B and C is 150°. If the force exerted by cord A is 100 N, determine graphically the values of the forces exerted by cords B and C, assuming the three forces to be in equilibrium.

3. A ship is being towed at constant speed by two tugs. The angles made by the tow-ropes with the direction of motion of the ship are 60° and 30° respectively. The force opposing the motion of the ship is 60 kN, acting along the line of its motion. Determine graphically the pull in each of the tow-ropes.

4. A vertical load of 100 N is supported by two chains, A and B, in the same vertical plane. The force in A is 50 N and acts in a line at 30° to the horizontal plane (i.e. at 120° to the 100-N load). Determine the force in chain B and the angle between A and B.

5. Two men carry a mass of 30 kg between them by means of two ropes fixed to the mass. One rope is inclined at 30° to the vertical and the other at 45°. Determine the tension in each rope.

6. A chain, 3 m long, has its ends fixed to a horizontal girder at points 2 m apart. A mass of 1 Mg is suspended from the chain at a point 1 m from one end. Determine the tensions in the two parts of the chain.

7. A mass of 30 kg is suspended by a cord attached to the ceiling, but is pulled to one side by a horizontal cord until the first cord makes an angle of 30° with the vertical. Draw a force diagram and determine the tensions in the two cords.

8. A pendant electric light fitting having a mass of 1 kg is to have its position adjusted by tying a cord to the flex and fastening the cord to a nail in a wall. If the cord makes an angle of 30° with the horizontal and the flex makes an angle of 60° with the horizontal, both measured above the horizontal, determine with the aid of a vector diagram the pull on the cord.

9. A chain AB is 2 m long and a hook is attached at a point C, 0·8 m from A. The ends A and B are attached to two points in the same horizontal line, 1·6 m apart. Determine the pull in each part of the chain when a mass of 2 t is suspended from the hook.

10. The jib and tie of a simple crane make angles of 60° and 30° respectively with the horizontal, both measured in an anticlockwise direction from the horizontal. Determine the force in each member when a mass of 5 t is suspended from the hook.

11. The vertical member of a jib crane is 3 m long. The jib has a length of 6 m and the length of the tie rod is 5 m. The compressive load in the jib is restricted to 50 kN. With the aid of a vector diagram drawn to scale, determine the maximum mass, in tonnes, that can be lifted by the crane and the corresponding tension in the tie rod.

12. A horizontal jib, 3 m long, is hinged to a vertical wall at its left-hand end. It is supported at its right-hand end by a stay wire making an angle of 60° with the horizontal, the upper end of the wire being secured to the same wall. If a mass of 5 t is hung on the jib, 2 m from the hinge, determine the magnitude and direction of the reaction at the hinge and the force in the stay wire.

13. A simple roof truss has a tie beam of 13 m and rafters of 10 m and 6 m. Taking the points as hinged, determine the forces in the three members of the truss when a vertical load of 6 kN is carried at the apex.

14. Determine the magnitude and nature of the forces in the three members of the pin-jointed frame shown below. The frame is simply supported at B and C, and carries a vertical load of 12 kN at A.

15. The figure on p. 38 shows a simple form of roof truss, freely supported at X and Y and carrying a vertical load of 2 kN at Z. Determine the reactions at X and Y and the force in each member, stating whether it is tensile or compressive.

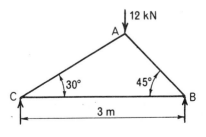

No. 14.– Examples 2.

16. Two forces of 8 N and 12 N act at a point, one force being inclined at an angle of 50° to the other. Determine the magnitude of the resultant force and its direction relative to the 8-N force.

17. The rope between a tug and the towed ship is inclined at an angle of 12 to the direction of travel of the ship. The tension in the rope is 500 N. Calculate the rectangular components of this tension, assuming one of the components to be in the direction of travel of the ship.

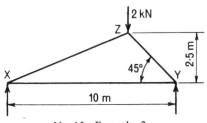

No. 15.– Examples 2.

18. A force of 30 N is inclined at an angle of 20° to the horizontal. Determine the component forces in the horizontal and vertical directions.

19. A truck is at rest on a level railway track. A horizontal pull of 250 N is applied at an angle of 30° with the direction of the rails. Calculate (a) the force tending to move the truck forward and (b) the sideways thrust on the rails.

20. A force of 20 N acts at an angle of 60° to a second force. If the resultant force is 28 N, determine graphically the magnitude of the second force.

21. The following horizontal forces act at a point:

> 50 N in direction due east,
> 80 N in direction due south,
> and 30 N in direction 20° north of west.

Calculate the value and the direction of a fourth force necessary to maintain equilibrium.

22. The following horizontal forces act at a point:

> 20 N in direction due north,
> 30 N in direction 20° south of east,
> 10 N in direction south-west,
> and 16 N in direction 5° south of west.

Calculate the value and direction of the resultant force. Check your value by means of a force polygon drawn to scale.

23. A load W is supported at a point O by three wires A, B and C, which act in the same vertical plane. The angles WOA, WOB and WOC, measured in a clockwise direction, are 120°, 210° and 240° respectively. With the aid of a force diagram drawn to scale, determine the value of W and the force in C if the forces in A and B are 20 N and 15 N respectively.

24. A ladder, 5 m long, rests at an angle of 60° to the horizontal with its upper end against a smooth vertical wall and its lower end on rough

ground. The ladder has a mass of 20 kg and the weight of the ladder may be assumed to act at a point 2 m from the lower end. Determine the reactions of the wall and of the ground on the ladder.

25. A uniform ladder, 4 m long and having a mass of 25 kg, rests against a smooth vertical wall at A and is supported on rough horizontal ground at B. The ladder is inclined at 60° to the horizontal. Determine the magnitude and direction of the reaction at the ground and at the wall respectively.

26. A ladder having a mass of 30 kg and a length of 6 m rests against a smooth vertical wall. The foot of the ladder stands on level ground 2·5 m away from the foot of the wall. If the whole weight of the ladder is assumed to act vertically downwards from a point one-third of its length from the end on the ground, determine the magnitude and direction of the reaction between (*a*) the ladder and the wall and (*b*) the ladder and the ground.

27. An overhanging horizontal platform AB is hinged at B and is supported by two chains attached to the outer edge and inclined 30° to the platform. The length AB is 4 m and the total weight of the platform and the load upon it is 10 kN and may be assumed to act at a distance of 3 m from the hinged side. Draw a force diagram to scale and determine the tension in each chain and the magnitude and direction of the reaction on the platform of the hinge at B.

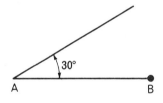

No. 27.– Examples 2.

28. A uniform steel bar is 4 m long and has a mass of 80 kg. Its lower end is hinged to a vertical wall and its upper end is supported by a horizontal chain, 2 m long, secured to a point in the wall directly above the hinge. Determine the tension in the chain and the value and direction of the reaction at the hinge.

29. A constant force of 40 N, acting at an angle of 20° with the horizontal, is applied to a body lying on a horizontal surface. Calculate the work done when the body is moved through a distance of 6 m along the surface. If the time taken is 3 s, what is the average value of the power?

ANSWERS TO EXAMPLES 2

1. 50 N, 53° below horizontal.
2. 153·2 N, 197 N.
3. 30 kN, 52 kN.

15. 0·5 kN at X, 1·5 kN at Y;
    1·58 kN compression in XZ,
    2·12 kN compression in YZ,

4. 86·6 N, 90°.
5. 216 N, 153 N.
6. 8·87 kN, 2·54 kN.
7. 340 N, 170 N.
8. 4·9 N.
9. 17·75 kN in AC,
   13·95 kN in BC.
10. Jib, 85 kN compression;
   tie, 49·05 kN tension.
11. 2·55 t, 41·7 kN.
12. 25 kN at 40·9° to jib,
   37·8 kN.
13. 5·6 kN, 4·2 kN, 3·8 kN.
14. 10·75 kN compression in AB,
   8·8 kN compression in AC,
   7·6 kN tension in BC.

1·5 kN tension in XY.
16. 18·2 N, 30·3°.
17. 489 N, 104 N.
18. 28·2 N, 10·26 N.
19. 216·5 N, 125 N.
20. 12 N.
21. 73 N, 72·6° north of west.
22. 5·36 N, 13·8° north of east.
23. 28·7 N, 11·35 N.
24. 45·4 N; 202 N, 77° to ground.
25. 255 N, 73·9° to ground; 70·7 N.
26. 45 N, normal to wall;
   298 N, 81·3° to ground.
27. 7·5 kN; 13·2 kN, 10·9°
   to horizontal.
28. 227 N; 818 N, 16·1° to vertical.
29. 225·6 J, 75·2 W.

CHAPTER 3

Moments and centre of gravity

3.1 Moment of a force about an axis

Suppose that a uniform strip or rod of wood, such as a metre rule, is pivoted at its mid-point C, as in fig. 3.1. If no external force is applied to the rod, there is no tendency for it to rotate either clockwise or anticlockwise about C.

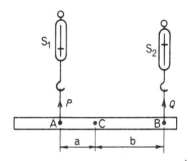

Fig. 3.1 Moments of parallel forces.

If we now attach the hook of a spring balance* S_1 to a cord passing through a hole at A, any upward force applied by this balance causes the rod to turn clockwise about pivot C. On the other hand, if we apply an upward force by a spring balance S_2 at point B (S_1 having been removed), the rod turns anticlockwise.

Let us next apply upward forces simultaneously by means of balances S_1 and S_2, of such values as to maintain the rod in a horizontal position. If P and Q be the forces exerted by S_1 and S_2 respectively, and if a and b be the distances of points A and B from pivot C, it is found that:

$$P \times a = Q \times b.$$

* The spring balances should be calibrated in newtons. The experiment can, however, be performed with known masses suspended from the rod; but in that case, it is the weight of each mass that must be used for calculating the moment. For example, if a mass of 3 kg is suspended from a horizontal rod at a distance of 0·6 m from the fulcrum, the weight of the mass $\simeq 3 \times 9·81 = 29·43$ N, and the moment about the fulcrum $= 29·43$ [N] $\times 0·6$ [m] $= 17·66$ N m.

This relationship can be confirmed by noting the balance readings for various upward forces exerted by the balances and the experiment can be repeated with various values of a and b.

The tendency of a force to produce rotation of a body about an axis is termed the *moment* of the force about that axis, and the magnitude of the moment is the product of the force and the perpendicular distance or *moment arm* of the line of action of the force from the axis. For instance, in fig. 3.1, if P is the force, in newtons, exerted by spring balance S_1 tending to turn the rod clockwise about an axis through C perpendicular to the paper, and if a is the distance, in metres, of the line of action of P from the axis through C, then:

clockwise moment of force P about this axis
$$= P \text{ [newtons]} \times a \text{ [metres]} = Pa \text{ newton metres.}$$

Similarly, anticlockwise moment of force Q about the same axis
$$= Qb \text{ newton metres.}$$

When the moments are such as to maintain the rod of fig. 3.1 horizontally, then:

clockwise moment due to P = anticlockwise moment due to Q
i.e. $Pa = Qb$

This relationship is termed the *principle of moments* and is discussed more generally in section 3.4.

3.2 Levers

A lever is simply a rod or bar capable of turning about a fixed axis called the *fulcrum*, which may be a spindle as in fig. 3.1 or a knife edge as in fig. 3.2. The lever may be straight or curved or cranked (section 3.7), and the forces acting upon it may be parallel or otherwise.

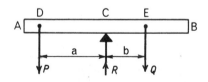

Fig. 3.2 A straight lever.

Let us first consider a straight uniform lever AB resting on a fulcrum at C, as in fig. 3.2. By arranging for the fulcrum to be midway along AB, we avoid any complication from the weight of the lever, since the gravitational pull on the right-hand half of the lever, tending to turn the lever clockwise about C, is balanced by that on the left-hand side tending to turn the lever anticlockwise. Consequently, when the lever is not loaded, there is no tendency for it to turn about the fulcrum C.

A downward force P, applied at a distance a from the fulcrum, gives rise to an anticlockwise moment Pa about C. Similarly, a downward force Q, applied at a distance b from the fulcrum, gives rise to a clockwise moment Qb about C. As already explained in section 3.1, the lever remains horizontal if the clockwise and anti-clockwise moments are equal,

i.e. if $$Pa = Qb.$$

The supporting or upward force R exerted by the fulcrum on the lever must be equal to the sum of the downward forces P and Q,

i.e. $$R = P + Q.$$

Note that we have taken moments about the axis through which the supporting force R acts on the lever. This force R has no moment about axis C and therefore does not affect the balance of the lever. Had moments been taken about any other axis, the moment of R about such an axis would have to be taken into consideration (see section 3.4).

Example 3.1 *A uniform lever (fig. 3.3) is pivoted at its mid-point C. A body having a mass of 5 kg is suspended at a point E, 180 mm to the right of C. Calculate the mass to be suspended at a point D, 400 mm to the left of C to maintain the lever in balance.*

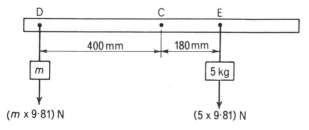

Fig. 3.3 Example 3.1.

Weight of the 5-kg mass $\simeq$ (5 $\times$ 9·81) N,

∴   clockwise moment of this force about C
$$= (5 \times 9\cdot81) \text{ [N]} \times 0\cdot18 \text{ [m]}.$$

If m be the mass, in kilograms, to be suspended at D,

weight of this mass $\simeq$ ($m \times 9\cdot81$) N

and its anticlockwise moment about C
$$= (m \times 9\cdot81) \text{ [N]} \times 0\cdot4 \text{ [m]}.$$

For balance, the clockwise and anticlockwise moments must be equal,

i.e. $5 \text{ [kg]} \times 9\cdot81 \text{ [N/kg]} \times 0\cdot18 \text{ [m]} = m \times 9\cdot81 \text{ [N/kg]} \times 0\cdot4 \text{ [m]}$

∴ $m = 2\cdot25 \text{ kg.}$

3.3 Equilibrant and resultant of parallel forces

In section 3.2 it was mentioned that the single upward force R exerted by the fulcrum on the lever, in fig. 3.2, opposes or balances the two downward forces P and Q. Force R is therefore the *equilibrant* of the other two forces. (The equilibrant and resultant of inclined forces have already been discussed in section 2.6.)

Let us consider the slightly more involved case shown in fig. 3.4, where three parallel downward forces of 80 N, 40 N and 20 N act on a light rod. It is required to determine the magnitude and position of a single vertical upward force (applied, say, by a spring balance or by a knife-edge support underneath the rod) that will balance these three downward forces.

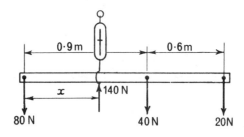

Fig. 3.4 Equilibrant of three parallel forces.

The upward force must obviously be equal to 80 + 40 + 20 = 140 N, as an experiment with a spring balance (fig. 3·4) would show; i.e. the *equilibrant* is 140 N. To determine the position of this equilibrant, let us take moments about an axis in the line of action of, say, the 80-N force. If x is the distance between the line of action of the equilibrant and that of the 80-N force, as shown in fig. 3.4, then:

anticlockwise moment of equilibrant about the chosen axis

$$= 140 \text{ [N]} \times x$$

and total clockwise moment of the other three forces about same axis

$$= 80 \text{ [N]} \times 0 \text{ [m]} + 40 \text{ [N]} \times 0 \cdot 9 \text{ [m]} + 20 \text{ [N]} \times 1 \cdot 5 \text{ [m]}$$

$$= 66 \text{ N m.}$$

Since the clockwise and anticlockwise moments are equal,

$$140 \text{ [N]} \times x = 66 \text{ [N m]}$$

$$\therefore \qquad\qquad x = 0 \cdot 471 \text{ m.}$$

Hence an upward force of 140 N in a line 0·471 m to the right of the 80-N force would balance the three downward forces. A *downward* force of 140 N in this line is the *resultant* of the three forces of 80 N, 40 N and 20 N. In other words, the three forces could be replaced by a single downward force of 140 N acting at a distance of 0·471 m from the line of action of the 80-N force. The resultant is exactly equal in magnitude and directly opposite to the equilibrant.

To take another example, suppose the 20-N force to be absent, i.e. consider only the downward forces 80 N and 40 N, 0·9 m apart, as shown in fig. 3.5. The magnitude of the equilibrant is 80 N + 40 N = 120 N. Suppose x to be the distance between the equilibrant and the 80-N force.

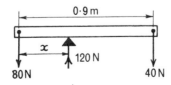

Fig. 3.5 Equilibrant of two parallel forces.

Taking moments about the line of action of the 80-N force, we have:

$$120 \ [\text{N}] \times x = 40 \ [\text{N}] \times 0\!\cdot\!9 \ [\text{m}]$$
$$\therefore \qquad x = 0\!\cdot\!3 \ \text{m}.$$

It will be seen that the line of action of the resultant (and of the equilibrant) divides the distance (0·9 m) between the 80-N and the 40-N forces *inversely* as the values of the forces, i.e. the distance from the 80-N force is 40/(80 + 40) of 0·9 m, namely 0·3 m, and that from the 40-N force is 80/120 of 0·9 m, namely 0·6 m.

3.4 General principle of moments

In the preceding sections we have taken moments about an axis on the line of action of one of the forces, for instance, about the fulcrum C in fig. 3.2. We shall now show that the Principle of Moments applies to *any* axis.

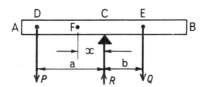

Fig. 3.6 General Principle of Moments.

Consider the arrangement shown in fig. 3.6 (similar to that of fig. 3.2), and let us take moments about *any* axis F situated at a distance x from the fulcrum C.

Anticlockwise moment of P about F $= P(a - x)$,

clockwise moment of Q about F $= Q(b + x)$

and   anticlockwise moment of R about F $= Rx$.

Equating the clockwise and anticlockwise moments, we have:

$$Q(b + x) = P(a - x) + Rx$$
Since $\qquad R = P + Q$
$$\therefore \qquad Qb + Qx = Pa - Px + Px + Qx$$
so that $\qquad Qb = Pa$

which is the condition of balance already derived in section 3.1. Hence we can express the General Principle of Moments thus:

If a body is at rest under the action of several forces, the total clockwise moment of the forces about **any** *axis is equal to the total anticlockwise moment of the forces about the same axis.*

If the forces are all in one plane, we can, for convenience in stating the principle, speak of moments about any *point* instead of about any *axis*.

3.5 Reactions on a horizontal beam supported at two points

A very common problem is to determine what upward forces must act on a beam at its supports when the beam is loaded and supported at two points. Often these two points are at its ends, but not necessarily so. Consider a uniform beam AB, 3 m long (fig. 3.7), resting on supports A and B at its ends. Suppose a body having a mass of 5 kg to be suspended at a point C, 2 m from A,

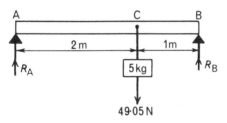

Fig. 3.7 Supporting forces at ends of a beam.

then downward force at C = weight of the 5-kg mass
$$\simeq 5 \times 9{\cdot}81 = 49{\cdot}05 \text{ N}.$$

The weight of the beam and of the suspended mass will exert downward forces on the supports at A and B; and the supports will exert exactly equal and opposite (upward) forces, or reactions, on the beam. These reactions are represented by R_A at A and R_B at B in fig. 3.7. For the present we shall ignore the weight of the beam, i.e. we shall regard the beam as of negligible weight or alternatively we shall regard the supporting forces as being the increases due to the load placed on the beam. In either case, it will be evident that the sum of the reactions R_A and R_B will be equal to the weight of the mass suspended at C,

i.e. $$R_A + R_B = 49{\cdot}05 \text{ N}.$$

The values of R_A and R_B can be determined by taking moments about any point along the lever; but the problem is simplified if we deal with only *one unknown force* at a time. This can be done by choosing an axis on one of the unknown reactions: for example, by choosing the axis through, say, A. The reaction R_A has no moment about A.

We can now imagine the beam of fig. 3.7 about to turn anticlockwise about A due to the action of the upward force R_B upon it. This effect, however, is exactly balanced by the action of the downward force of 49·05 N at C. In other words, the moment due to R_B tending to turn the beam anticlockwise about A is balanced by the moment due to the force at C tending to turn the beam clockwise about A.

Hence, $R_B \times 3 \ [\text{m}] = 49\cdot05 \ [\text{N}] \times 2 \ [\text{m}]$

$\therefore$ $R_B = 32\cdot7 \ \text{N}.$

Similarly, if we had chosen B as our axis, the clockwise moment due to reaction R_A would be $R_A \times 3 \ [\text{m}]$, and the anticlockwise moment due to the force at C would be $49\cdot05 \ [\text{N}] \times 1 \ [\text{m}]$.

Hence, $R_A \times 3 \ [\text{m}] = 49\cdot05 \ [\text{N m}]$

$\therefore$ $R_A = 16\cdot35 \ \text{N}.$

Alternatively, since $R_A + R_B = 49\cdot05 \ \text{N}$

$\therefore$ $R_A = 49\cdot05 - 32\cdot7 = 16\cdot35 \ \text{N}.$

Or, *after* having determined the value of R_B, we could have considered rotation about C, i.e. we could have taken moments about C thus:

$$R_A \times 2 \ [\text{m}] = 32\cdot7 \ [\text{N}] \times 1 \ [\text{m}]$$

$\therefore$ $R_A = 16\cdot35 \ \text{N}.$

Effect of weight of the beam. The effect of the weight of the beam on the downward forces on the supports and on the equal upward forces of the supports on the beam is obviously to increase them by an amount equal to the weight of the beam. Thus, if the mass of the beam in fig. 3·7 is, say, $0\cdot2 \ \text{kg}$, its weight is approximately $0\cdot2 \times 9\cdot81$, namely $1\cdot96 \ \text{N}$; and since the beam is uniform, each reaction is increased by $0\cdot98 \ \text{N}$,

i.e. $R_A = 16\cdot35 + 0\cdot98 = 17\cdot33 \ \text{N}$

and $R_B = 32\cdot7 + 0\cdot98 = 33\cdot68 \ \text{N}.$

If the supports are not symmetrically placed with respect to the centre of a uniform beam, the two reactions due to the weight of the beam will not be equal. For the purpose of these simple problems in statics, we may take the weight of the beam (although actually distributed) as acting at a point in the beam called its *centre of gravity*. We shall defer more general consideration of centre of gravity until a little later (section 3.10), but for the present we may be satisfied by the fact that the centre of gravity of a straight uniform rod is midway between its ends and the weight of the rod may be taken as acting there.

Example 3.2 *A uniform horizontal beam, 6 m long, rests on two supports A and B, 4 m apart, A being at one end of the beam. The mass of the beam is 20 kg. Calculate the reactions of the supports on the beam.*

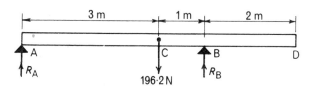

Fig. 3.8 Example 3.2.

Let R_A and R_B be the reactions at A and B respectively, as shown in fig. 3.8.

The gravitational force on the beam, i.e. the weight of the beam
$$\simeq 20 \times 9\cdot81 = 196\cdot2 \text{ N}.$$

Taking moments about A, we have:

clockwise moment $= 196\cdot2 \text{ [N]} \times 3 \text{ [m]} = 588\cdot6 \text{ N m}$

and   anticlockwise moment $= R_B \times 4 \text{ [m]}.$

Equating the clockwise and anticlockwise moments, we have:

$$R_B \times 4 \text{ [m]} = 588\cdot6 \text{ [N m]}$$
$$\therefore \qquad R_B = 147\cdot15 \text{ N}.$$

Total upward reactions $= R_A + R_B$
$$= \text{weight of beam} = 196\cdot2 \text{ N},$$
$$\therefore \qquad R_A = 196\cdot2 - 147\cdot15 = 49\cdot05 \text{ N}.$$

Alternatively, taking moments about B, we have:

$$R_A \times 4 \text{ [m]} = 196 \cdot 2 \text{ [N]} \times 1 \text{ [m]}$$

$\therefore$
$$R_A = 49 \cdot 05 \text{ N}$$

and
$$R_B = 196 \cdot 2 - 49 \cdot 05 = 147 \cdot 15 \text{ N.}$$

Example 3.3 *If a mass of 8 kg be hung from the beam of Example 3.2 at a distance of 1 m from A, calculate the reaction of the supports.*

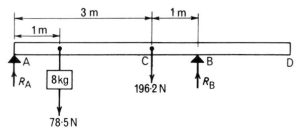

Fig. 3.9 Example 3.3.

Weight of the 8-kg mass $\simeq 8 \times 9 \cdot 81 = 78 \cdot 5$ N.

We proceed as in Example 3.2, but this time we have to take account of the two downward forces of 78·5 N and 196·2 N (fig. 3.9). To find reaction R_B, we take moments about A thus:

$$\text{clockwise moment} = 78 \cdot 5 \text{ [N]} \times 1 \text{ [m]} + 196 \cdot 2 \text{ [N]} \times 3 \text{ [m]}$$
$$= 78 \cdot 5 + 588 \cdot 6 = 667 \cdot 1 \text{ N m.}$$

Anticlockwise moment $= R_B \times 4$ [m].

Equating clockwise and anticlockwise moments, we have:

$$R_B \times 4 \text{ [m]} = 667 \cdot 1 \text{ [N m]}$$

$\therefore$
$$R_B = 166 \cdot 8 \text{ N}$$

and
$$R_A = 196 \cdot 2 + 78 \cdot 5 - 166 \cdot 8 = 274 \cdot 7 - 166 \cdot 8$$
$$= 107 \cdot 9 \text{ N.}$$

Alternatively, taking moments about B, we have:

$$R_A \times 4 \text{ [m]} = 78 \cdot 5 \text{ [N]} \times 3 \text{ [m]} + 196 \cdot 2 \text{ [N]} \times 1 \text{ [m]}$$
$$= 431 \cdot 7 \text{ N m}$$

$\therefore$
$$R_A = 107 \cdot 9 \text{ N}$$

and
$$R_B = 274 \cdot 7 - 107 \cdot 9 = 166 \cdot 8 \text{ N.}$$

Example 3.4 *If a mass of 6 kg be suspended at the end of the projecting beam of Example 3.2, calculate the reactions at the supports.*

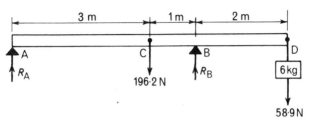

Fig. 3.10 Example 3.4.

Weight of the 6-kg mass $\simeq 6 \times 9\cdot81 = 58\cdot9$ N.

Taking moments about A (fig. 3.10), we have:

clockwise moment $= 196\cdot2$ [N] $\times 3$ [m] $+ 58\cdot9$ [N] $\times 6$ [m]
$= 942$ N m.

Anticlockwise moment $= R_B \times 4$ [m]

Hence $\qquad R_B \times 4$ [m] $= 942$ [N m]

$\therefore \qquad\qquad\qquad R_B = 235\cdot5$ N

and $\qquad\qquad R_A = 196\cdot2 + 58\cdot9 - 235\cdot5 = 255\cdot1 - 235\cdot5$
$= 19\cdot6$ N.

Alternatively, taking moments about B, we have:

$R_A \times 4$ [m] $+ 58\cdot9$ [N] $\times 2$ [m] $= 196\cdot2$ [N] $\times 1$ [m]

$\therefore \qquad\qquad\qquad R_A = 19\cdot6$ N

and $\qquad\qquad R_B = 255\cdot1 - 19\cdot6 = 235\cdot5$ N.

Example 3.5 *If the mass suspended at the projecting end in Example 3.4 is increased to 12 kg, calculate the reactions of the supports.*

Weight of the 12-kg mass $\simeq 12 \times 9\cdot81 = 117\cdot7$ N.

Taking moments about A (fig. 3.10), we have:

$196\cdot2$ [N] $\times 3$ [m] $+ 117\cdot7$ [N] $\times 6$ [m] $= R_B \times 4$ [m]

$\therefore \qquad\qquad\qquad R_B = 323\cdot7$ N

and $\qquad\qquad R_A = 196\cdot2 + 117\cdot7 - 323\cdot7$
$= -9\cdot8$ N.

c

This means 9·8 N in the opposite direction to the assumed upward direction, i.e. 9·8 N *downward*.

Alternatively, taking moments about B, we have:

$$R_A \times 4 \, [m] + 117·7 \, [N] \times 2 \, [m] = 196·2 \, [N] \times 1 \, [m]$$

$$\therefore \qquad R_A = -9·8 \, N; \text{ or } 9·8 \, N \text{ downward,}$$

$$\text{and} \qquad R_B = 196·2 + 117·7 + 9·8$$

$$= 323·7 \, N, \text{ as before.}$$

3.6 Other forms of straight levers

At (*a*), (*b*) and (*c*) in fig. 3.11 are shown the three possible forms of straight lever in which an effort *F* may overcome the force due to a load *W* (both forces being perpendicular to the lever) by turning the lever about a fulcrum C.

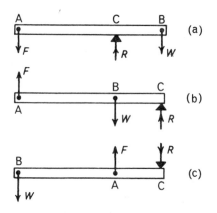

Fig. 3.11 Forms of straight lever.

Case (*a*) was dealt with in section 3.2, and examples of this class of lever are: beam balances in which the arms may be of equal lengths, as in a science laboratory balance, or of unequal lengths, as in a steelyard—a device consisting of a mass which can be moved along a graduated beam to balance a relatively heavy body suspended near to the fulcrum; a seesaw, etc.

Case (*b*) is, in principle, identical with the beam carrying a load as dealt with in section 3.5; and examples of this class of lever are: a wheelbarrow, a nut cracker, etc.

Case (*c*) represents a large effort *F* moving through a small distance and overcoming a small load W which moves through a large distance. For example, the manipulation of a fishing rod; the large upward pull of the biceps muscle acting about the elbow as fulcrum and lifting a relatively small load held in the hand.

The principle involved is the same in each of the above cases; and by equating opposing moments of a known load W and an unknown effort F about the fulcrum C, we can determine the value of F. Also, if desired, we can find the reaction R of the fulcrum C on the lever by taking moments about A or by finding what force is required to balance the sum or the difference of F and W, according to whether they act in the same or in opposite directions.

3.7 Bell crank lever

A lever is not necessarily straight, but may be cranked as in the bell crank lever shown in fig. 3.12. The value of a force P required to balance a force Q, neither of which need be perpendicular to the arm of the lever to which it is applied, can be found by considering

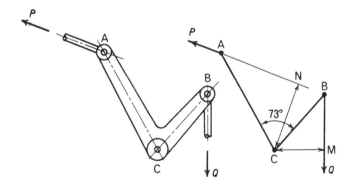

Fig. 3.12 Bell crank lever.

the opposing moments of P and Q about the axis C. The moment arms of P and Q about the axis C may readily be found by drawing the skeleton or centre-line diagram to scale as shown on the right-hand side of fig. 3.12. They are CN and CM respectively. Equating the anticlockwise moment of P about C to the clockwise moment of Q, we have:

$$P \times \text{distance CN} = Q \times \text{distance CM}.$$

Example 3.6 *In fig. 3.12, the length AC is* 180 *mm and that of BC is* 120 *mm, and angle ACB is* 73°. *Measuring the perpendicular distances of the lines of action of P and Q from axis C, we find CN =* 115 *mm and CM =* 82 *mm. If Q is* 60 *N, calculate the value of force P required to maintain a balance.*

Taking moments about C, we have:

$$\text{clockwise moment} = 60\ [\text{N}] \times 0{\cdot}082\ [\text{m}]$$
$$= 4{\cdot}92\ \text{N m}$$

and   anticlockwise moment = $P \times 0{\cdot}115$ [m].

Equating the clockwise and anticlockwise moments, we have:

$$P \times 0{\cdot}115\ [\text{m}] = 4{\cdot}92\ [\text{N m}]$$
$$\therefore \qquad P = 42{\cdot}8\ \text{N}.$$

3.8 Application of the principle of moments

Many problems in statics, such as finding the value of an unknown force acting on a body at rest when its line of action is known, can easily be solved by this principle.

Two points are of great importance to the beginner. The first is to decide upon which body or part of a body or structure we are going to concentrate attention. In the case of simple levers or beams already considered, there was only one body and therefore no doubt about the body on which the various forces were exerted (though it was necessary to take the forces exerted *on* the lever and not those equal and opposite forces exerted *by* the lever or beam on the supports). But even in a simple structure, not only is the structure as a whole at rest under the action of forces but also its several parts are at rest, and we can apply the principle to the whole or to any part as may be most convenient.

The second point to be observed is that the point or axis about which we take the equal and opposite moments should, if possible, be chosen so that the moments include that of only *one* unknown force. This is generally done by choosing a point on the line of action of another unknown force if there are more than one. The method can best be understood by considering a few examples.

Example 3.7 *A wall crane is represented diagrammatically by the centre lines of its members in fig. 3.13 and has a load of 3 kN at O. Determine the tension in the tie-rod QR. This tie-rod is freely jointed at Q and R.*

The unknown tension T in the tie-rod RQ exerts a pull at the ends R and Q. Member PQ is at rest and the forces exerted upon it are: (*a*) the load of 3 kN, (*b*) the unknown tension T (which must act along the tie-rod QR) and (*c*) whatever force may be exerted upon PQ through the pivot or hinge at the end P. This last force, like T, is an unknown one. In order that it shall not have any effect on any equation of moments, consider the forces tending to cause rotation and their moments about an axis perpendicular to the diagram and *through* P. Then any reaction or external force acting at P has no moment about this axis.

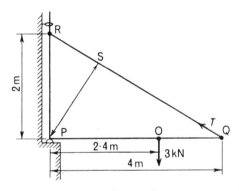

Fig. 3.13 Wall crane diagram

On drawing the diagram to scale or by calculation, we find that the distance PS, which is the moment arm of the tension T about the axis selected, is 1·79 m. Hence,

anticlockwise moment about P $= T \times 1\cdot79$ [m]
and clockwise moment about P $= 3000$ [N] $\times 2\cdot4$ [m] $= 7200$ N m.

Equating the clockwise and anticlockwise moments, we have:

$$T \times 1\cdot79 \text{ [m]} = 7200 \text{ [N m]}$$
$$\therefore \qquad T = 4020 \text{ N} = 4\cdot02 \text{ kN}.$$

This problem could also have been solved graphically by means of a force diagram drawn to scale, as already explained for Example 2.2.

Example 3.8 *The simple triangular roof frame or truss shown in fig. 3.14(a) carries a load of 12 kN at its top joint. Determine the forces in the members AB, BC and AC.*

Since the frame is symmetrical about a vertical axis through B, the vertical upward supporting forces at A and C are each 6 kN. To find the force which BC exerts on the joint at C, consider the fact that the tie-rod AC is at rest and the forces acting on it are in

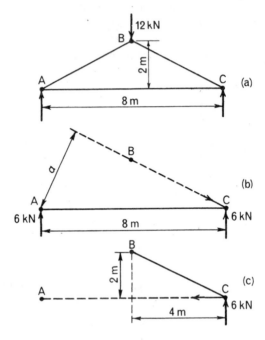

Fig. 3.14 Roof frame diagram

equilibrium, as shown in fig. 3.14(*b*). Then the moment tending to turn AC anticlockwise about an axis through A and perpendicular to the plane of the diagram

$$= \text{vertical reaction at C} \times \text{distance AC}$$
$$= 6000 \text{ [N]} \times 8 \text{ [m]} = 48\,000 \text{ N m.}$$

Moment tending to turn AC clockwise about A

$$= \text{thrust of BC on joint C} \times \text{distance } a.$$

From a scaled diagram, distance *a* is found to be 3·58 m. Hence:

$$\text{thrust of BC} \times 3·58 \text{ [m]} = 48\,000 \text{ [N m]}$$
$$\therefore \qquad\qquad \text{thrust of BC} = 13\,400 \text{ N} = 13·4 \text{ kN.}$$

This is the compressive force in the member BC; and by symmetry, the compressive force in the member BA must be the same.

In order to find the force in member AC, note that it exerts a force on C and therefore tends to rotate BC about an axis through B, perpendicular to the figure, as shown in fig. 3.14(c). Considering the forces acting on BC and having moment about B, we have:

moment tending to turn BC anticlockwise about B

> = vertical reaction at C × *horizontal* distance of this reaction from B

= 6000 [N] × 4 [m] = 24 000 N m,

and moment tending to turn BC clockwise about B

> = horizontal pull of AC × vertical distance from B to AC
> = pull of AC × 2 [m].

Since these moments balance each other,

$$\text{pull of AC} \times 2 \text{ [m]} = 24\ 000 \text{ [N m]}$$
∴ $$\text{pull of AC} = 12\ 000 \text{ N} = 12 \text{ kN}.$$

This is the tension in member AC, i.e. the pull which it exerts on the joints at C and A, and is equal to the pulls or reactions which the joints exert on the tie-rod AC. The arrowheads in fig. 3.14 (*b*) and (*c*) indicate the directions of the forces exerted *by* the members *on* the hinges or pin-joints, *not* the forces exerted by the joints *on* the members.

3.9 Couples

Fig. 3.15 shows two equal forces *F* acting on a body such as a beam AB. The forces are in opposite directions but *their lines of action are not the same*. The only effect of these forces is to produce a turning effect.

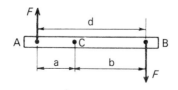

Fig. 3.15 A couple.

Let us consider any point C in the plane of the two forces, and suppose the distances between C and the lines of action of the upward and downward forces to be a and b respectively. Then:

clockwise moment about an axis through C due to the force acting upward $= Fa$.

Similarly the clockwise moment about the same axis due to the force acting downward $= Fb$,

$\therefore$ total clockwise moment about an axis through C

$$= Fa + Fb = Fd$$

where $d = a + b$

= perpendicular distance between the lines of action of the two forces.

Two equal parallel forces acting in opposite directions, as in fig. 3.15, are termed a *couple*; and the *moment of a couple* is the product of one of the forces and the perpendicular distance between the lines of action of the forces, i.e. Fd in fig. 3.15.

A couple cannot be balanced by a *single* force; it can only be balanced by a couple of equal moment acting in the opposite direction.

One example of a couple is the tightening or loosening of a wing nut. Another example, shown in fig. 3.16(a), consists of equal forces F applied to two cords attached to the periphery of a drum (or wheel) of diameter d. As already explained, the moment of the

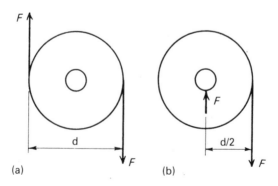

Fig. 3.16 A couple (a) with no reaction at support and (b) with reaction at support.

couple is *Fd*. In this case, the couple does not exert any force on the bearings supporting the shaft to which the drum is attached. (There will, of course, be an upward force exerted by the bearings to balance the weight of the drum and shaft, but this force is independent of the value of forces *F*.)

Let us next consider the case of a drum (or wheel) with only *one* cord attached to its periphery, as shown in fig. 3.16(*b*). When a downward force *F* is applied to the cord, the bearings must exert an equal upward force *F*, and the two forces form a couple having a moment = $F \times d/2$, where *d* is the diameter of the drum.

Well-known examples of this type of couple are the tightening or loosening of a nut by means of a spanner and the winding of a rope on a drum or a capstan.

3.10 Centre of gravity

Every particle of matter is attracted towards the centre of the earth and every object or body consists of a large number of particles. If the body is small compared with the earth, the gravitational forces on all the particles of the body can, for practical purposes, be regarded as being parallel with one another. These parallel forces can be replaced by a resultant force equal to the weight of the body and having its line of action passing through a point termed the *centre of gravity* (abbreviation, c.g.) of the body. In other words, the mass of a body may be regarded as being concentrated at its centre of gravity and its weight acts vertically through that point. The location of the centre of gravity is independent of the position of the body. If a body is suspended by a cord, the vertical upward

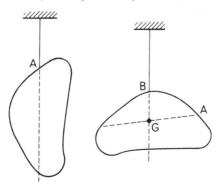

Fig. 3.17 Determination of the centre of gravity of a thin plate by suspension.

force exerted by the cord must be equal to the weight of the body and its line of action passes through the centre of gravity of that body.

It is often possible to determine the position of the centre of gravity by a simple experiment. For instance, fig. 3.17 shows a thin plate of irregular shape. If the plate is suspended in turn from two points, A and B, and if vertical lines from A and B respectively are drawn as shown dotted in fig. 3.17, the intersection G of these lines gives the position of the centre of gravity of the plate.

In the case of each of the following symmetrically-shaped homogeneous solids, the centre of gravity is at its geometrical centre. The plates are assumed to be of uniform thickness so that their centres of gravity are at mid-thickness.

Rectangular plate: c.g. at intersection of diagonals.

Triangular plate: c.g. at intersection of *medians* or lines joining an apex to the mid-point of the opposite side. It is one-third of the distance from the mid-point towards the apex.

Circular plate: c.g. at centre.

Sphere: c.g. at centre.

Cylinder: c.g. at mid-point of the axis.

Ring: c.g. at centre (which is not in the body of the ring).

We shall now consider some examples of bodies of less symmetrical shape which are divisible into parts. If we know the centre of gravity of each part and the relative *weights* of the parts, we can determine the centre of gravity by the principle of moments.

Example 3.9 *Determine the centre of gravity of the T-shaped piece of uniform sheet metal shown in fig. 3.18.*

Let w be the weight of the sheet in newtons per square millimetre.

Area of cross-piece ABCD $= 60 \,[\text{mm}] \times 20 \,[\text{mm}] = 1200 \,\text{mm}^2$

$\therefore$ weight of cross-piece ABCD $= 1200 \, w$ newtons.

Area of stem KLFE $= 80 \,[\text{mm}] \times 20 \,[\text{mm}] = 1600 \,\text{mm}^2$

$\therefore$ weight of stem KLFE $= 1600 \, w$ newtons,

and   total weight of sheet $= 1200 \, w + 1600 \, w$

$= 2800 \, w$ newtons.

The c.g. of the cross-piece ABCD is at G in the centre line and is 10 mm from AB.

The c.g. of the stem KLFE is at H in the centre line and is $20 + 40 = 60$ mm from AB.

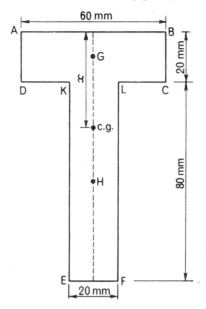

Fig. 3.18 Example 3.9.

If we take moments about AB as axis, the moment of the whole piece about AB must be equal to the sum of the moments of the two parts about AB. Hence, if x be the distance of the c.g. from AB,

$$2800 \ w \ [\text{N}] \times x = 1200 \ w \ [\text{N}] \times 10 \ [\text{mm}] + 1600 \ w \ [\text{N}] \times 60 \ [\text{mm}]$$
$$\therefore \qquad\qquad x = 38 \cdot 6 \text{ mm},$$

i.e. the c.g. is in line GH at a distance of 38·6 mm from AB.

Example 3.10 *Determine the c.g. of the uniform rectangular plate, 320 mm × 160 mm, with a circular hole, 120 mm diameter, cut out in the position shown in fig. 3.19.*

Let w be the weight of the plate in newtons per square millimetre.

$$\text{Total weight of plate with no hole} = w \times 320 \times 160$$
$$= 51 \ 200 \ w \text{ newtons.}$$
$$\text{Area of circular plate removed} = (\pi/4) \times 120 \times 120$$
$$= 11 \ 300 \text{ mm}^2$$

∴   weight of circular plate removed $= w \times 11\,300$ newtons,
and weight of remaining perforated plate $= 51\,200\,w - 11\,300\,w$
$$= 39\,900\,w \text{ newtons.}$$

Let x be the distance of the c.g. of the perforated plate from axis AD.

The c.g. of the unperforated plate is at O, 160 mm ($= 0.16$ m) from AD, and that of the circular piece is 240 mm ($= 0.24$ m) from AD.

Taking moments about axis AD, we have:

moment due to weight of perforated plate $= 39\,900\,w\,[\text{N}] \times x$

moment due to weight of circular plate $= 11\,300\,w\,[\text{N}] \times 0.24\,[\text{m}]$
$$= 2712\,w\,[\text{N m}]$$
and moment due to weight of whole plate $= 51\,200\,w\,[\text{N}] \times 0.16\,[\text{m}]$
$$= 8192\,w\,[\text{N m}].$$

Equating the sum of the moments of the parts to the moment of the whole rectangular plate, we have:

$$39\,900\,w\,[\text{N}] \times x + 2712\,w\,[\text{N m}] = 8192\,w\,[\text{N m}]$$
∴
$$x = 0.1373\,\text{m}$$
$$= 137.3\,\text{mm.}$$

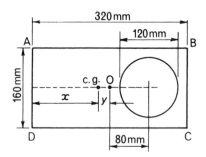

Fig. 3.19 Example 3.10.

Alternatively, and as a check, we can determine the distance of the c.g. of the perforated plate from the centre O by taking moments about an axis through O and parallel to AD.

Suppose y to be the distance of the c.g. of the perforated plate from O. This c.g. must lie to the left of O.

The moment due to the weight of the unperforated plate about O is zero. Hence the clockwise moment due to the weight of the circular piece must balance the anticlockwise moment of the perforated plate,

i.e. $11\ 300\ w$ [N] $\times$ 0·08 [m] $=$ 39 900 w [N] $\times\ y$

∴ $y = 0·0227$ m

                        $= 22·7$ mm,

which corresponds to $160 - 22·7 = 137·3$ mm from AD.

3.11 Stable, unstable and neutral equilibrium

A body is said to be in *stable* equilibrium when, if slightly displaced from its position, the forces acting upon it tend to cause it to return to that position. If, however, the forces acting upon it, after a slight displacement, tend to move it further from its original position, the equilibrium is said to be *unstable*. On the other hand, if, after a displacement of the body, the forces acting upon it remain in equilibrium, the body tends neither to return to its original position nor to be displaced further. The equilibrium is then said to be *neutral*.

Let us consider a rectangular iron plate of uniform thickness, having its centre of gravity at C and supported by a rod passing through a hole A near one end of the plate, as shown in fig. 3.20.

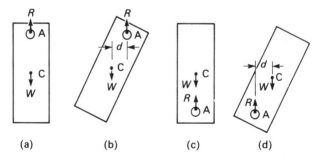

(a)   (b)   (c)   (d)

Fig. 3.20 Stable and unstable equilibrium

If the supporting rod is horizontal, the plate takes up the position shown in fig. 3.20(a) such that the c.g. of the plate is *directly* below the supporting rod; i.e. the reaction R of the rod on the plate is in line with and equal to the weight W of the plate.

If the plate is moved slightly clockwise so that the distance between the lines of action of R and W is d, as in fig. 3.20(b), there is a couple Wd (see section 3.9), acting anticlockwise, tending to return the plate to its original position shown in fig. 3.20(a), namely the position of *stable* equilibrium.

If the plate is turned so that its c.g. is *exactly above* the supporting rod, as in fig. 3.20(c), forces R and W are in line with each other, and the plate could theoretically remain in that position. If, however, the plate were slightly displaced, say, clockwise to the position shown in fig. 3.20(d), there would be a couple Wd tending to turn the plate further clockwise, until it would ultimately take up the stable position of fig. 3.20(a). Hence fig. 3.20(c) shows the plate in a position of *unstable* equilibrium.

Let us next consider a solid sphere of iron, such as a ball-bearing, resting on a *perfectly smooth horizontal* surface S, as shown in fig. 3.21

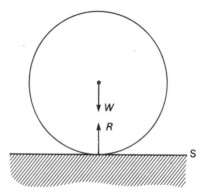

Fig. 3.21 Neutral equilibrium

The weight W of the sphere and the reaction R of the supporting surface are exactly equal and opposite for every position of the sphere. Consequently, if the sphere is displaced in any direction, there is no couple available to move it either backward or forward, and the equilibrium is said to be *neutral*.

Summary of Chapter 3

The moment of a force about an axis is the product of the force and the perpendicular distance of the line of action of the force from the axis.

General principle of moments. If a body is at rest under the action of several forces, the total clockwise moment of the forces about any axis is equal to the anticlockwise moment of the forces about the same axis.

The principle of moments applies to parallel and non-parallel forces. It also applies to cranked levers and to framed structures, the joints of which can be assumed to act as frictionless hinges.

In the case of *parallel* forces, the principle of moments can be used to determine the magnitude and position of a single force, called the *equilibrant*, to balance several parallel forces. This force, reversed, is the *resultant* of the several parallel forces. An important application is the location of the resultant of the distributed weight of a body. It always acts through the centre of gravity of the body.

EXAMPLES 3

1. A wooden rod, whose weight may be neglected, is pivoted at a point A. A spring balance is attached vertically to the rod at a point B, 200 mm from A, and another spring balance is similarly attached to the rod at a point C on the other side of A, the distance between C and A being 360 mm. What must be the reading on the balance at C when the balance at B reads 80 N, in order that the rod may remain in equilibrium? What is the downward force exerted by the pivot on the rod?

2. Where must the fulcrum be situated in a straight lever, 2 m long, if a downward effort of 200 N at one end is to lift a mass of 80 kg at the other end? State the distance from the effort.

3. A lever AB, 3 m long, rests on a fulcrum at C, 1 m from A. There is a downward force of 80 N acting on the lever at A. Calculate (*a*) the force required at B to maintain equilibrium and (*b*) the force exerted by the fulcrum on the lever. Neglect the weight of the lever.

4. A lever AB, 2 m long, is pivoted at end A. A mass of 3 kg is suspended at a point C, 0·8 m from A. Calculate the vertical force required at B to maintain the lever horizontally. Neglect the weight of the lever.

5. If the mass of 3 kg in Q. 4 were suspended at point B of the lever, what would be the vertical force required at point C to maintain the lever horizontally?

6. A rod AB of uniform section, 2 m long, has a mass of 5 kg. The rod rests on a fulcrum 0·6 m from A. Calculate (*a*) the vertical force required at B to maintain the rod in a horizontal position and (*b*) the value of the reaction at the fulcrum.

7. A uniform horizontal lever AE, 7 m long, is supported on a fulcrum at C, 3 m from A. There are downward forces of 20 N and 30 N at A and E respectively and another downward force of 10 N at a point B, 2 m from A. What vertical force must be applied at a point D, 1 m from E, in order that the lever may remain horizontal? Will this force act upward or downward? Neglect the weight of the lever.

8. A horizontal steel beam of uniform section, 4 m long, rests on two supports, one at the end of the beam and the second at a distance of 1 m from the other end of the beam. The mass of the beam is 80 kg. Calculate the upward reactions at the supports.

9. If a 50-kg mass is suspended midway between the supports of the beam in Q. 8, calculate the reactions of the supports.

10. If a 50-kg mass is suspended half-way along the overhanging length of the beam in Q. 8, calculate the reactions of the supports.

11. The mass of a uniform steel girder, 12 m long, is 3 Mg. It is supported at one end and at a point 5 m from the other end. Calculate the reactions of the two supports.

12. If a mass of 2 Mg is suspended at the end of the overhanging portion of the beam in Q. 11, calculate the reactions of the supports.

13. If the 2-Mg mass in Q. 12 were suspended from the beam midway between the supports, what would be the reactions of the supports?

14. A uniform horizontal beam, 6 m long, is supported at points A and B, 1 m and 4 m respectively, from one end. If the supporting force at B is 400 N, calculate the mass of the beam and the supporting force at A.

15. If a uniform beam, 5 m long, is supported 1 m from one end, how far from that end must a second support be placed in order that it shall support 60 per cent of the weight of the beam?

16. Two men carry a scaffolding pole, 10 m long, having a mass of 50 kg, each supporting one end. The centre of gravity of the pole is 4 m from one end. How much weight does each man support?

   If a bag of tools having a mass of 15 kg is to be carried by slinging it over the pole, at what distance from the lighter end of the pole should it be slung in order that the total load should be shared equally?

17. A beam AB is 15 m long and is freely supported at A and B. Loads of 60 kN, 100 kN and 80 kN are placed on the beam at distances of 3 m, 6 m and x metres from A. Calculate the distance x if the loads on the supports are equal. Neglect the weight of the beam.

18. A uniform beam AB has a mass of 1 Mg and is 4 m long. It is supported at A and B. A mass of 3 Mg is suspended at C, 1 m from A, and a rope exerting an upward pull of 5 kN is attached to the beam at D, 1·5 m from B. Calculate the reactions at A and B.

19. A square plate ABCD, of 100 mm side, is free to rotate in a vertical plane about a peg through its centre. Forces of 8 N, 20 N, 12 N and 16 N act along AB, CB, DC and DA respectively. Calculate the magnitude and sense of the resultant turning moment about the peg. What would be the length of the arm of a couple composed of 5-N forces which would maintain the plate in equilibrium?

20. In order to find the position of the centre of gravity of a small connecting rod, it is suspended in a horizontal position from two spring balances, one at each end. The reading on the balance at the big end is 32 N and that on the other balance is 12 N. If the distance between the points of support is 180 mm, determine the distance of the centre of gravity from the big end.

21. A cranked lever ACB has a horizontal arm AC, 400 mm long, and an arm

CB, 150 mm long. The lever can turn freely about a pin at C and the angle ACB is 120°. What vertical pull will be required at A to produce a horizontal pull of 200 N in a wire attached at B?

22. A uniform bar, 2 m long, has a mass of 25 kg. It is hinged at one end and is kept in a horizontal position by a rope attached to the free end, making an angle of 45° with the bar and in the same vertical plane as the bar. If a mass of 60 kg is hung on the bar 1·2 m from the hinge, what is the tension in the rope?

23. A wall crane has the dimensions shown in the diagram and a mass of 500 kg is suspended at C. Determine the forces in AC and BC, and state whether they are tensile or compressive.

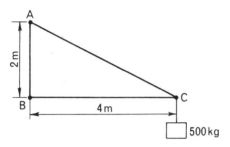

No. 23.– Examples 3.

24. A simple roof truss has the dimensions shown in the diagram and a load of 12 kN at A. Determine the forces in the members of the frame and state which are in tension and which are in compression.

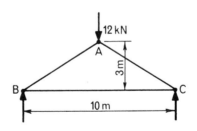

No. 24.– Examples 3.

25. A simple pin-jointed structure is in the form of an isosceles triangle ABC in which A is the apex and the supports are at B and C. If AB = AC = 5 m, and BC = 8 m, determine the force in each member when a vertical load of 4 kN is applied at A.

26. A metal plate of uniform thickness is in the form of an equilateral triangle

of 2-m sides. If the plate has a mass of 200 kg/m² and is lying flat on the ground, calculate the value of the vertical force to be applied at one corner of the plate in order to lift it to an upright position by rotating it about the opposite side (as about a hinge).

27. A triangular plate ABC of wrought iron, 10 mm thick, has the following dimensions: AB, 2 m; BC, 3 m; CA, 2 m. The plate lies flat on a horizontal surface. What vertical force must be applied to the corner A in order to lift this corner? Density of wrought iron is 7 800 kg/m³.

28. A T-shaped piece of metal of uniform thickness has a cross-piece 100 mm × 30 mm and a stem 140 mm × 25 mm. Determine the distance of the centre of gravity from the top edge of the T-piece.

29. An L-shaped piece of steel plate of uniform thickness can be divided into two rectangles, each 60 mm × 20 mm. Determine the distances of the centre of gravity from the left-hand and bottom edges respectively of the L-piece.

30. A rectangle, 90 mm × 60 mm, is cut from a thin rectangular plate, 100 mm × 70 mm, to form an L-section, the width of the legs being 10 mm. Determine the centre of gravity of the L-section.

31. A sheet of metal, 400 mm square, has centre lines drawn on it parallel to the sides. A circle, 250 mm diameter, is cut out of the square, the centre of the circle lying on one of the centre lines of the square. Determine the distance between the centres of the square and of the circle if the centre of gravity of the remaining plate is 30 mm from the centre of the square.

32. A hole having a diameter of 80 mm is cut out from a uniform plate of 250 mm diameter. The distance between the centre of the plate and the centre of the hole is 50 mm. Calculate the position of the centre of gravity of the remaining material.

ANSWERS TO EXAMPLES 3

1. 44·4 N, 124·4 N.
2. 1·593 m.
3. 40 N, 120 N.
4. 11·77 N.
5. 73·6 N.
6. 14 N, 35·05 N.
7. 16·7 N, upward.
8. 261·6 N, 523·2 N.
9. 506·8 N, 768·5 N.
10. 179·8 N, 1095·5 N.
11. 4·2 kN, 25·23 kN.
12. −9·81 kN, 58·86 kN.
13. 14·02 kN, 35·03 kN.
14. 61·2 kg, 200 N.
15. 3·5 m.
16. 294·3 N, 196·2 N; 1·665 m.
17. 12·75 m.

18. 25·1 kN, 9·14 kN.
19. 0·4 N m, 40 mm radius.
20. 49·1 mm.
21. 64·95 N.
22. 673 N.
23. 11 kN tension in AC, 9·81 kN compression in BC.
24. 11·66 kN compression in AB and AC, 10 kN tension in BC.
25. 3·33 kN compression in AB and AC, 2·67 kN tension in BC.
26. 1132 N.
27. 506 N.
28. 60·8 mm.
29. $x = 20$ mm, $y = 30$ mm.
30. $x = 18·1$ mm, $y = 33·2$ mm.
31. 67·9 mm.
32. 5·7 mm from centre.

CHAPTER 4

Inclined plane

4.1 Smooth inclined plane

The use of the inclined plane for lifting heavy bodies by means of small forces has been known since ancient times; for instance, the large stone blocks forming the Pyramids were lifted in this way.

In practice, it is not possible to make an inclined plane perfectly smooth, but much the same effect is obtained by the use of rollers or wheels under the body to be hauled up the plane.

It is instructive to employ more than one method to determine the force required to hold a body in equilibrium on an inclined plane or to haul it up the plane at a constant speed, and to see that all the methods lead to the same conclusions. The inclined plane thus forms an excellent example to illustrate matters dealt with in chapters 1, 2 and 3.

4.2 Effort parallel to inclined plane

Fig. 4.1(a) represents a metal roller attached by a cord to a spring

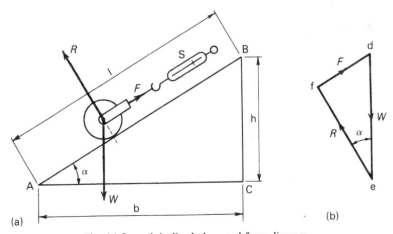

Fig. 4.1 Smooth inclined plane and force diagram.

balance S. The roller rests on a smooth surface AB inclined at an angle α to the horizontal. The weight W of the roller can be determined by suspending it from a spring balance calibrated in newtons.

Suppose the reading on spring balance S to be F newtons when the roller is stationary or is being hauled up the inclined surface at a steady speed.

The contact between a perfect cylindrical surface and a perfectly flat surface is a straight line parallel to the axis of the cylinder. Since the inclined surface AB in fig. 4.1 is being assumed smooth, the direction of the reaction R between the cylinder and that surface is represented by a straight line drawn at right angles to the surface. and passing through the line of contact and also through the axis of the cylinder, as shown in fig. 4.1(a).

The relationship between F and W can be derived by several methods, utilizing principles discussed in earlier chapters.

(a) By force diagram

The three forces F, W and R, acting on the body, are in equilibrium and can therefore be represented by the triangular force diagram shown in fig. 4.1(b), where vectors de, ef and fd are parallel to W, R and F respectively, and the lengths of the vectors represent to scale the magnitudes of the respective forces.

Hence,
$$\frac{fd}{de} = \frac{F}{W}$$

or
$$F = \frac{fd}{de} \times W.$$

Since the line of action of W is at right angles to AC and that of R is at right angles to AB, the angle between the directions of W and R must be equal to angle α between AC and AB. Also, the angle between the lines of action of forces F and R is 90°. Hence the triangle formed by fd, de and ef in fig. 4.1(b) is similar to triangle ABC in fig. 4.1(a), so that:

$$\frac{fd}{de} = \frac{BC}{AB}$$

∴
$$F = \frac{fd}{de} \times W = \frac{BC}{AB} \times W$$
$$= \frac{h}{l} \times W = W \sin α \qquad (4.1)$$

where $\quad l =$ length of plane

$\qquad h =$ height of plane

and $\quad \alpha =$ angle between the inclined plane and the horizontal.

(b) By moments

If we choose any point O on the line of action of R, fig. 4.2, distant ON from the centre of gravity of the roller; then the anticlockwise moment of F about an axis through O, perpendicular to the diagram, must balance the clockwise moment of W about the same

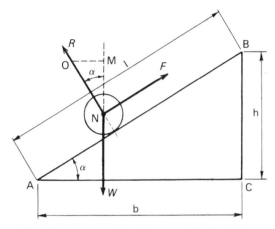

Fig. 4.2 Moment solution for smooth inclined plane.

axis (the moment of R about axis O is zero),

hence $\qquad\qquad F \times ON = W \times \mathrm{OM}$

$\therefore \qquad\qquad\qquad F = \dfrac{\mathrm{OM}}{ON} \times W.$

Since triangle NOM is similar to triangle ABC,

$\therefore \qquad\qquad\qquad \dfrac{\mathrm{OM}}{ON} = \dfrac{\mathrm{BC}}{\mathrm{AB}}$

so that $F = \dfrac{\mathrm{OM}}{ON} \times W = \dfrac{\mathrm{BC}}{\mathrm{AB}} \times W = \dfrac{h}{l} \times W = W \sin \alpha$, as before.

(c) By resolution into components

Let us resolve force W (fig. 4.1) into two components, one downward along the plane and the other perpendicular to the plane. The component downward along the plane is:

$$W \cos \text{ABC} = W \times h/l.$$

This component, being the only component acting downward along the plane, must balance force F acting upward along the plane, (all other forces being perpendicular to the plane);

hence, $F = W \times h/l = W \sin \alpha$, as before.

The other component of W, namely $W \cos \alpha$, must be equal in magnitude and opposite in direction to reaction R.

(*d*)*By equating the work done by effort to the work done on load*

If the friction at the bearings of the roller in fig. 4.1(a) is negligible, then the work done by the effort F in hauling the roller up the inclined plane is all expended in lifting the roller. Thus, if the roller is hauled a distance l metres along the inclined plane by an effort F newtons acting parallel to the plane,

work done *by* effort $= Fl$ joules.

If h metres be the height through which the roller has been raised, and if the weight of the roller is W newtons,

work done in lifting the roller $= Wh$ joules.

Since the effect of friction is being assumed negligible,

work done by effort $=$ work done in lifting the roller

i.e. $Fl = Wh$

∴ $F = W \times h/l = W \sin \alpha$, as before.

The work done by or against reaction R is zero since the direction of R is at right angles to the direction of motion of the roller.

4.3 Effort horizontal

Suppose the roller to be hauled up the inclined plane at a constant speed by a horizontal effort F, as shown in fig. 4.3(a). The three forces, F, W and R, can be represented by the triangular force diagram of fig. 4.3(b). From this diagram it is seen that:

$$\frac{F}{W} = \frac{fd}{de} = \tan \alpha$$

∴ $F = W \tan \alpha = W \times h/b$ (4.2)

where b is the distance through which F moves *in its own direction* while the roller is raised through a distance h.

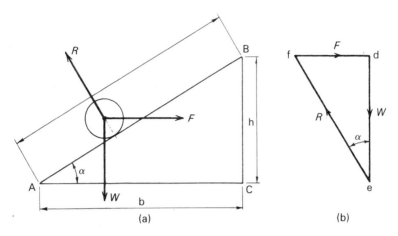

Fig. 4.3 Horizontal effort on smooth inclined plane.

Alternatively,

$$\text{work done by effort } F = Fb$$
and work done in lifting roller $= Wh$.

Equating the work done by the effort to the work done in lifting the roller, we have:

$$Fb = Wh$$
∴ $$F = W \times h/b = W \tan \alpha, \text{ as before.}$$

Reaction R in fig. 4.3(a) does no work since its direction is perpendicular to the direction of motion of the roller.

4.4 Effort at any angle

If the effort F is applied at any angle θ to the inclined plane, as in fig. 4.4(a), to haul the roller up the plane at a constant speed, the value of F can be found by drawing the vector diagram shown in fig. 4.4(b). Vector *de* is drawn vertically to represent W to scale, and from d and e, lines are drawn parallel to F and R respectively, to meet at f. The value of F is represented to scale by the length of vector *fd*.

An expression for the value of effort F can be derived thus:

component of effort P acting along plane $= P \cos \theta$.

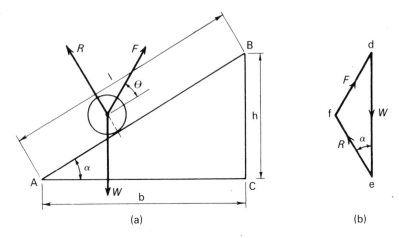

(a)   (b)

Fig. 4.4 Effort at any angle on smooth inclined plane.

Work done by this component in hauling the roller through a distance l along the inclined plane $= (F \cos \theta) \times l$.

If the roller is raised through a height h,

work done by effort F = work done in lifting roller

Since the reaction R is at right angles to the direction of motion of the roller,

work done on or by reaction $R = 0$.

Since friction is assumed to be negligible,

work done by effort P = work done in lifting roller

i.e. $(F \cos \theta) \times l = W \times h$

$\therefore$
$$F = \frac{W}{\cos \theta} \times \frac{h}{l}$$
$$= W \times \frac{\sin \alpha}{\cos \theta} \qquad (4.3)$$

The same result can be obtained by resolving F and W along the inclined surface, thus:

component of F acting up the incline $= F \cos \theta$

and   component of W acting down the incline $= W \sin \alpha$.

Since the component of R acting parallel to the inclined plane is zero, the components of F and W must be equal and opposite,

hence $\qquad\qquad\qquad F \cos \theta = W \sin \alpha$

$\therefore \qquad\qquad\qquad\qquad F = W \dfrac{\sin \alpha}{\cos \theta}$, as before.

Example 4.1 *A smooth inclined plane has a gradient of 1 in 4. Calculate the force required to haul a mass of 10 kg up the gradient at a constant speed when the direction of the force is* (a) *parallel to the plane,* (b) *horizontal and* (c) *at an angle of 20° with the plane.*

A gradient of 1 in 4 means that there is a rise of 1 m for every 4 m travelled up the inclined plane.

Weight of the 10-kg mass $= W \simeq 10 \times 9\cdot81$
$$= 98\cdot1 \text{ N.}$$

(*a*) If the mass is hauled a distance of 4 m along the inclined plane by an effort F parallel to the plane,

work done by effort $= F \times 4$ [m].

During this time the mass has been raised to a height of 1 m above its original position against a gravitational force (or weight) of $98\cdot1$ N,

$\therefore \qquad$ work done on the mass $= 98\cdot1$ [N] $\times$ 1 [m].

Since the plane is assumed to be smooth,

work done by the effort $=$ work done on the mass

i.e. $\qquad\qquad F \times 4$ [m] $= 98\cdot1$ [N] $\times$ 1 [m]

$\therefore \qquad\qquad\qquad\qquad F = 24\cdot5$ N.

(*b*) Let b be the horizontal base corresponding to a 4-m length of plane, then:

$$4^2 = b^2 + 1^2$$

$\therefore \qquad\qquad\qquad b^2 = 15$

so that $\qquad\qquad\qquad b = 3\cdot87$ m

Hence, while the horizontal effort F travels 3.87 m, the mass is lifted 1 m. Equating the work done by the effort to the work done in lifting the mass, we have;

$$F \times 3.87 \, [\text{m}] = 98.1 \, [\text{N}] \times 1 \, [\text{m}]$$
$$\therefore \qquad\qquad F = 25.4 \, \text{N}.$$

(c)   Component of F along the plane $= F \cos 20°$
$$= F \times 0.94,$$

∴ work done by effort in hauling the mass through a distance of 4 m along the inclined plane

$$= (F \times 0.94) \times 4 \, [\text{m}].$$

During this time the mass has been lifted 1 m,

$\therefore$ work done on the mass $= 98.1 \, [\text{N}] \times 1 \, [\text{m}].$

Equating the work done by the effort to the work done on the mass, we have:

$$(F \times 0.94) \times 4 \, [\text{m}] = 98.1 \, [\text{N}] \times 1 \, [\text{m}]$$
$$\therefore \qquad\qquad F = 26.1 \, \text{N}.$$

Example 4.2 *A train having a mass of* 300 *t is hauled up a gradient of* 1 *in* 250. *The tractive resistance on the level is* 4.5 *mN/N (or* 4.5 *N/kN), i.e. for every newton of train weight, the tractive force required to overcome wind resistance and rail and axle friction is* 4.5 *mN. Calculate the tractive effort required to haul the train at a constant speed. Also, calculate the tractive effort required to haul the train down the gradient at a constant speed.*

$$\text{Mass of train} = 300 \, \text{t}$$
$$= 300 \, 000 \, \text{kg},$$
$\therefore$ weight of train $\simeq 300 \, 000 \times 9.81$
$$= 2.943 \times 10^6 \, \text{N}$$
$$= \text{vertical force on track.}$$

Since the gradient is only 1 in 250, the force perpendicular to the track is practically the same as the vertical force on the track, namely 2.943×10^6 N.

Hence, effort to overcome $\Big\}$ = $(2{\cdot}943 \times 10^6)$ [N] $\times$ 4·5 [mN/N]
tractive resistance

$$= 13{\cdot}25 \times 10^6 \text{ mN}$$
$$= 13\ 250 \text{ N.}$$

From expression (4.1),

effort to haul train up gradient if there were no friction
$$= (2{\cdot}943 \times 10^6) \text{ [N]} \times 1/250 = 11\ 770 \text{ N.}$$
$\therefore$ total tractive effort = 13 250 + 11 770 = 25 020 N
$$= 25{\cdot}02 \text{ kN.}$$

When the train is being hauled down the gradient, gravitation, in the form of the component of the weight of the train along the gradient, is providing a force of 11 770 N towards overcoming the tractive resistance of 13 250 N opposing motion.

Hence the force required to $\Big\}$ = 13 250 − 11 770 = 1480 N
haul the train down the gradient
$$= 1{\cdot}48 \text{ kN.}$$

Summary of Chapter 4

By the use of an inclined plane, a heavy body can be lifted by a relatively small force. The ratio of the effort to the weight of the body lifted can be found by: (a) the force diagram, (b) moments, (c) the resolution of forces acting on the body and (d) equating the work done by the effort to the work done in lifting the body.

When the effort is along the inclined plane,

$$F = W \sin \alpha \qquad (4.1)$$

When the effort is horizontal,

$$F = W \tan \alpha \qquad (4.2)$$

When the effort makes an angle θ with the plane,

$$F = W \frac{\sin \alpha}{\cos \theta} \qquad (4.3)$$

EXAMPLES 4

1. A body having a weight of 500 N is placed on a smooth inclined plane, 5 m long. The top of the plane is 2 m above the lower end. Calculate the

force required to hold the body in equilibrium if it is applied (*a*) parallel
to the plane, (*b*) horizontally.

2. A body having a mass of 20 kg is pulled at a uniform speed up a smooth
inclined plane having a gradient of 1 in 3. Calculate the force required if
it acts (*a*) parallel to the plane, (*b*) horizontally.

3. A smooth inclined plane has a gradient of 1 in 6. Calculate the mass of
a body that can be held in equilibrium on the plane by a force of 80 N
when the force acts (*a*) parallel to the plane, (*b*) horizontally.

4. A body having a mass of 5 t is hauled at a constant speed up a smooth
plane which is inclined at 10° to the horizontal. Calculate the value of the
force required parallel to the plane.

5. An inclined plane rises 1 m for every *x* metres of its length. A force of
60 N, acting parallel to the plane, is required to haul a body having a
weight of 450 N at a uniform speed up the plane. Calculate the value of
x, assuming friction resistance to be negligible.

6. A body having a mass of 30 kg rests on a smooth plane inclined at 20°
to the horizontal and held in equilibrium by a horizontal force *P*. Sketch
the force diagram and determine graphically the magnitudes of force *P*
and of the reaction between the load and the inclined plane.

7. A body having a weight of 60 N rests on a smooth plane inclined at 30°
to the horizontal. Determine (*a*) the horizontal force required to prevent
the body from sliding down the plane, (*b*) the value and direction of the
least force required to pull the body up the plane.

8. A smooth inclined plane has a gradient of 1 in 5. A body having a mass
of 60 kg rests on the plane. Draw the force diagram to scale and, from
the diagram, determine the effort required to pull the body at a uniform
speed up the plane if the effort makes an angle of 30° with the plane.

9. The slope of a smooth inclined plane is 20° to the horizontal. A force of
400 N, inclined at an angle of 15° to the plane, holds a body in equilibrium.
Determine, graphically or otherwise, the mass of the body.

10. A train having a mass of 400 t travels at a constant speed along a straight
track. At that speed, the tractive force required to overcome friction and
wind resistance is 5 mN/N (or 5 N/kN). Calculate the tractive effort
required to maintain the same speed: (*a*) on a level track, (*b*) up an incline
of 1 in 150 and (*c*) down an incline of 1 in 300.

11. A train is hauled up an incline of 1 in 40 by a tractive effort of 20 kN
acting parallel to the slope. Calculate the distance travelled and the work
done, in megajoules, when the train has been raised through a vertical
distance of 30 m.

ANSWERS TO EXAMPLES 4

1. 200 N, 218 N.
2. 65·4 N, 69·3 N.
3. 48·9 kg, 48·2 kg.
4. 8·51 kN.
5. 7·5 m.
6. 107 N, 313 N.

7. 34·6 N, 30 N parallel to plane.
8. 136 N.
9. 115 kg.
10. 19·62 kN, 45·78 kN, 6·54 kN.
11. 1·2 km, 24 MJ.

CHAPTER 5

Friction

5.1 Friction

The term 'friction' has been referred to in earlier chapters and its general meaning is widely understood, but here we consider its effects as a definite force affecting problems of equilibrium.

Suppose a body to be resting on a horizontal table. It is acted upon by a downward force W due to the gravitational pull of the earth upon it (i.e. its weight). There must also be a reaction N normal to the surface of the table. This reaction, exerted *by* the table *on* the body, is equal and opposite to force W, as shown in fig. 5.1, where—for clarity—the line of action of N is shown displaced slightly from that of W. Actually, *they are in direct opposition.*

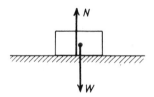

Fig. 5.1 Normal forces only.

If the body were resting on a surface inclined to the horizontal, the value of N would be less than that of W and the line of action of N would not be directly opposite that of W.

If a horizontal force P is applied to the body, pulling towards the right as in fig. 5.2, the body will not move unless P is large enough. This is due to the presence of a force of friction F exerted on the body *by the table* towards the left and just sufficient to balance force P. This force of friction F has the feature of *adjustment* so as to be exactly equal and opposite to P, i.e. if P is increased, F increases by exactly the same amount, until motion occurs.

While the body is at rest, we may regard it as being in equilibrium under the action of the *four* forces shown in fig. 5.2. The table top

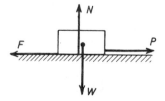

Fig. 5.2 Frictional drag.

exerts *two* forces, N vertically (equal and opposite to W) and F horizontally (equal and opposite to P), as shown in fig. 5.3(a). From the force diagram of fig. 5.3(b), it is seen that these two forces can

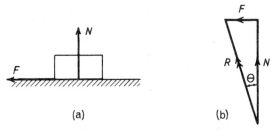

Fig. 5.3 Resultant reaction.

be replaced by a resultant force R^* acting at an angle θ to the left of the line of action of force N, the value of θ being given by:

$$\tan \theta = F/N \qquad (5.1)$$

We may therefore regard the body as being in equilibrium under the action of *three* forces, namely W, P and R, shown in fig. 5.4(a). These forces can be represented vectorially by the triangular force diagram of fig. 5.4(b).

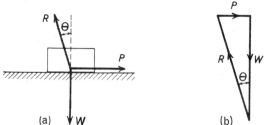

Fig. 5.4 Three forces acting on body.

* The double arrowhead on R indicates that R represents the resultant of two forces and *not* the equilibrant of those forces.

As the pulling force P is increased, the friction force F (equal and opposite to P) also increases and the resultant R inclines more to the left, i.e. angle θ increases. There is, however, a *limit* to this adjustment of F to resist a growing value of P. This limit is reached when the body is on the point of motion towards the right. Thus, when motion just begins to take place, the friction operating against motion is the maximum or *limiting* friction.

This condition (i.e. when component F has reached its limiting value, or what is the same thing, when R has reached its greatest inclination to the vertical) is shown in fig. 5.5(a). The corresponding angle between R and the perpendicular or normal to the sliding surface is termed the *angle of friction* and is represented by φ.

From expression (5.1) it follows that:

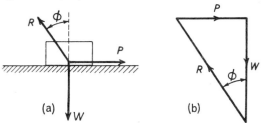

Fig. 5.5 Angle of friction.

$$\tan \varphi = \frac{\text{sliding friction force } F}{\text{normal reaction } N}$$

The ratio of the sliding friction force to the normal reaction of the supporting surface is termed the *coefficient of static friction* when the object is just on the point of sliding and is represented by μ,

i.e. $$\mu = \frac{\text{sliding friction force}}{\text{normal reaction}} = \tan \varphi \qquad (5.2)$$

So far we have referred to just the limiting frictional force that occurs when an object is just on the point of sliding, such friction being referred to as static friction since the object is not moving. However, friction still occurs when an object is sliding, the friction in this situation being called dynamic friction. The frictional force in such a situation is generally slightly less than the limiting frictional force for the static condition. We thus have a coefficient of dynamic friction, defined in the same way as for static friction, which is slightly less than the static coefficient.

For a body, of weight W, pulled along a *horizontal* surface by a *horizontal* force P, P and W are equal and opposite to the sliding friction force and the normal reaction respectively; hence for this condition:

$$\mu = P/W \qquad (5.3)$$

Expression (5.3) is not applicable if the pulling force P is inclined to the surface or if the surface is not horizontal (see Example 5.2).

5.2 Determination of the coefficient of friction

Fig. 5.6 shows a simple apparatus by which experiments on the amount of friction between two surfaces can be performed. A wooden board A is supported horizontally on a table or bench; and on A rests a rectangular block B of known weight.

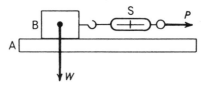

Fig. 5.6 Experiment on friction.

A spring balance S, calibrated in newtons, is attached to B. The horizontal pull P applied to S is increased until B begins to move, and the reading on S is noted: (*a*) just before the movement begins and (*b*) while B is being pulled along at a steady speed. As mentioned in section 5.1, the force to overcome *dynamic friction* is slightly less than that required to overcome *static friction*. The test is repeated with various values of load W by placing metal blocks of known weight on top of block B.

Experiments between pairs of solids with *dry* surfaces show that if the surfaces are of uniform character throughout in regard to finish and condition, the force P required to maintain steady slow motion is approximately proportional to the perpendicular force W between the two sliding surfaces, i.e. the ratio P/W, namely the coefficient of friction, is approximately constant.

5.3 Other experimental results

It is also found from experiment that for *dry* solid surfaces the frictional force.

(*a*) depends upon the nature of the surfaces in contact;

(*b*) is independent of the area of the surfaces in contact;

(*c*) is independent of the speed of sliding.

The above must be regarded as only rough empirical rules applicable in regard to relatively slow motion and moderate pressures and not necessarily correct at high speeds or high pressures or in regard to any but dry surfaces. Actually it is very difficult to exclude some form of lubrication, however slight its amount may be. Consequently experiments on friction are difficult and the results are not simple.

Example 5.1 *A block of metal having a mass of* 60 *kg requires a horizontal force of* 140 *N to drag it at a constant speed along a horizontal floor. Calculate (a) the coefficient of friction and (b) the angle of friction.*

$$\text{Since mass of block} = 60 \text{ kg.}$$

$$\therefore \quad \text{weight of block} \simeq 60 \times 9\cdot81 = 588\cdot6 \text{ N.}$$

$$\text{Friction force} = 140 \text{ N,}$$

$$\therefore \quad \text{coefficient of friction} = \frac{140 \quad [\text{N}]}{588\cdot6 \ [\text{N}]} = 0\cdot238$$

$$= \tan \varphi$$

From trigonometrical tables, $\varphi = 13\cdot4°$.

Example 5.2 *Determine the force, inclined at an angle of* 20° *to the horizontal, required to drag the* 60-*kg metal block of Example* 5.1 *along the floor at a uniform speed.*

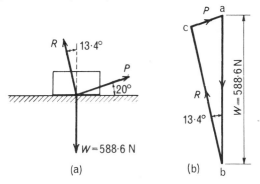

Fig. 5.7 Diagrams for Example 5.2.

D

Fig. 5.7(a) represents the three forces P, W and R acting on the metal block. The triangular force diagram of Fig. 5.7(b) may be constructed by drawing the vertical line ab 117·7 mm long to represent the weight W to scale of 1 mm to 5 N. Line bc is then drawn to represent R, making an angle of 13·4° with ba, where 13·4° is the angle of friction φ already determined in Ex. 5.1. Line ac is drawn parallel to force P to form the force triangle abc.

From the diagram it is found that the length of line ac representing P is 27·3 mm. Hence P = 27·3 [mm] × 5 [N/mm] = 136·5 N.

It will be seen that the value of P is slightly less than when it is acting horizontally. The value is a minimum when P is acting at right angles to R, i.e. when P is inclined at the angle of friction (13·4°) to the horizontal.

Example 5.3. *A block of stone having a mass of 75 kg is hauled along a horizontal floor for a distance of 100 m in 2 min. The coefficient of friction is 0·3. Calculate (a) the horizontal force required, (b) the work done and (c) the power.*

$$\text{Weight of block} \simeq 75 \times 9\cdot81 = 736 \text{ N,}$$
$$\therefore \quad \text{horizontal force required} = 736 \times 0\cdot3 = 220\cdot8 \text{ N.}$$

$$\text{Work done} = 220\cdot8 \text{ [N]} \times 100 \text{ [m]}$$
$$= 22\,080 \text{ J} = 22\cdot08 \text{ kJ,}$$
$$\text{and power} = \frac{22\,080 \text{ [J]}}{(2 \times 60) \text{ [s]}} = 184 \text{ W.}$$

5.4 Lubrication

Sometimes friction force is desirable as, for instance, in the grip of a tyre on a road, or a belt on a pulley or a brake block on a drum. But in machinery, where one piece has sliding or turning motion on or in another, any friction force acting against the motion results in loss of energy which is converted into heat. It is therefore desirable to reduce friction, and this is done by interposing a lubricant between the solid surfaces, thereby keeping them out of direct contact with each other.

Oils of various kinds are the best known lubricants; but for heavy loads, greases of various degrees of viscosity, which can resist pressure and avoid being squeezed out, are used. The lubricant employed depends upon many circumstances, such as the pressure in a bearing, and the speed as well as the cost. For high speeds, forced lubrication is used in which oil is supplied to bearings under pressure from a pump, the ideal condition being that the metal surfaces shall be completely separated by a film of oil in which a rotating shaft floats.

The resistance offered to lateral motion by a liquid is quite unlike that between solid surfaces. It increases with speed and with area of contact, and pressure has little or no effect on it. In practice friction resistance is often a complicated matter of forces exerted by solids and intervening fluids, on which not very much light is thrown by simple experiments on the sliding resistance between solid bodies, and even this is affected by variable amounts of films of grease, moisture, or even air partially separating the surfaces.

Summary of Chapter 5

Friction is a force opposing the sliding of one body over another. It acts at the surface of contact in a direction opposite to that of motion.

For a horizontal surface,

$$\text{coefficient of friction} = \mu = P/W \qquad (5.3)$$
$$= \tan \varphi$$

where $P =$ horizontal force to overcome friction force,

$W =$ weight of body

and $\varphi =$ angle of friction.

The force required to overcome sliding friction is slightly less than that required to overcome static friction.

The friction between one body and another body sliding over it can be reduced by separating the two surfaces by means of a lubricant.

EXAMPLES 5

1. A sledge having a mass of 400 kg requires a horizontal force of 620 N to keep it moving at a uniform speed along a horizontal surface. Calculate (a) the coefficient of friction and (b) the direction and magnitude of the resultant reaction of the surface.
2. A 2-t mass rests on a horizontal surface. Calculate the horizontal force, in kilonewtons, required to move this mass at a uniform speed along the surface if the coefficient of friction is 0·35.

3. A body having a mass of 40 kg rests on a horizontal table and it is found that the least horizontal force required to move the body is 90 N. What is the coefficient of friction?

   How much work is done in moving the body along the surface of the table through a distance of 3 m?

4. A wooden tray rests on a horizontal surface and it is found that with a mass of 3 kg in the tray, a horizontal force of 10 N is just sufficient to cause the tray to slide. When the 3-kg mass is replaced by a mass of 7 kg, it is found that the horizontal force required to cause the tray to slide is doubled. Calculate the coefficient of friction between the tray and the surface, and the mass of the tray.

   What would be the work done in moving the tray through a distance of 5 m with an 11-kg mass in the tray?

5. A block having a weight of 300 N rests on a flat horizontal surface, the coefficient of friction between the block and the surface being 0·25. Calculate the magnitude of the horizontal force which will just cause the block to slide over the surface.

   If the block slides in the direction of the applied force with a uniform velocity of 1·5 m/s, calculate the heat generated by friction, in joules per minute.

6. The sliding face of a slide valve of a steam engine is 150 mm by 300 mm, and the steam pressure on the back of the valve is 1200 kN/m². If the coefficient of friction is 0·02, what is the force required to move the valve?

7. A certain man has a mass of 70 kg. What is the magnitude of the largest mass he can pull by a horizontal rope along a horizontal floor if the coefficient of friction between the mass and the floor is 0·23 and that between his boot soles and the floor is 0·5?

8. A block of stone is hauled along a horizontal floor by a force inclined 20° to the horizontal. If the block has a mass of 40 kg and the coefficient of friction between the stone and the floor is 0·3, determine graphically or otherwise the effort required.

9. A mass of 30 kg resting on a rough horizontal surface is moved at a constant speed by a force of 40 N acting at 30° to the horizontal. Determine the coefficient of friction.

10. A body having a weight of 150 N is resting on a horizontal table and can just be moved by a horizontal force of 25 N. Calculate the coefficient of friction and the direction and magnitude of the resultant reaction of the table.

    If the force were acting at an angle of 20° with the horizontal, what would be the value of the force that could just move the body and what would be the resultant reaction of the table?

11. A mass of 12 t is dragged over a rough horizontal surface at a uniform speed. The load is moved a distance of 50 m in 12 min. If the coefficient of friction is 0·4, determine (a) the horizontal force required, (b) the work done, in kilojoules and (c) the power, in kilowatts.

ANSWERS TO EXAMPLES 5

1. 0·158, 3970 N, 9° with vertical.
2. 6·86 kN.
3. 0·229, 270 J
4. 0·255, 1 kg, 150 J.
5. 75 N, 6750 J/min.
6. 1080 N.

7. 152 kg.
8. 113 N.
9. 0·126.
10. 0·167, 152 N, 9·5° with vertical;
    25·2 N, 143·5 N.
11. 47·1 kN, 2355 kJ, 3·27 kW.

88

CHAPTER 6

Velocity and acceleration

6.1 Displacement

When we say that a person has walked 7 kilometres, we give no
indication either of the actual distance between his initial and final
positions or of the direction of the final position relative to the
initial position. He could be at any point within a radius of 7 km:
he could even be back at his initial position. *Distance* is therefore
a scalar quantity, i.e. it has magnitude only. If, however, the person
has walked 3 km eastwards, as represented by AB in fig. 6.1, and
then walked 4 km northwards, as represented by BC, his final
position C is 5 km away from his initial position A. This *change of
position* is termed the *displacement* and is independent of the path

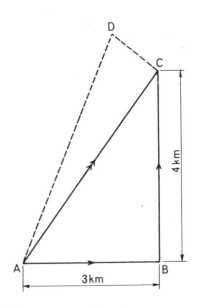

Fig. 6.1 Displacement.

followed and of the time taken; thus, he could have reached C by going north-north-east from A to D and then south-east from D to C, as shown by the dotted lines in fig. 6.1.

Since displacement has both magnitude and direction, it is a *vector* quantity and can be represented by a straight line drawn to scale in the direction of the displacement. From fig. 6.1, it will be be seen that during the first part of the 7-km journey from A to C, the displacement is 3 km in an easterly direction and this can be represented by a horizontal vector AB, drawn to a scale of, say, 1 cm to 1 km. During the second part of the journey, the displacement is 4 km in a northerly direction and is represented in fig. 6.1 by vector BC, 4 cm long, drawn vertically at B.

The resultant displacement is represented to scale by the straight line AC. By measurement, the length of AC is found to be 5 cm and the angle between AC and AB is 53°; i.e. the resultant displacement is 5 km in a direction 53° north of east and can be determined by adding vectorially the component displacements.

6.2 Speed

If a motor car travels 1·6 km in 2 minutes, its speed is 0·8 kilometres per minute, or $0·8 \times 60 = 48$ kilometres per hour. In general, we can say that the speed of a body is the distance traversed in unit time, or the *rate* at which distance is traversed. The distance can be expressed in any convenient unit such as a metre, a kilometre, etc., and the unit of time can also be any convenient value such as an hour, a minute or a second.

If a motor car travels a distance of 48 km in one hour, its *average* speed is 48 km/h, but it is extremely unlikely that the car will travel at exactly this speed during the whole hour—its speed will be at times higher and at other times lower than this value. A body has *constant* speed only if it moves over equal distances in equal intervals of time—however short the interval.

The average speed of a body is the total distance divided by the time; thus, if a body travels a distance s metres in t seconds, the average speed, v metres per second, is given by:

$$v = \frac{s \text{ [metres]}}{t \text{ [seconds]}} = \frac{s}{t} \text{ metres/second} \qquad (6.1)$$

Example 6.1 *If a motor car is travelling at a speed of* 100 *km/h, what is the speed in metres/second?*

$$\text{Since } 1 \text{ km} = 1000 \text{ m}$$
$$\text{and } 1 \text{ h} = 3600 \text{ s,}$$

$\therefore$ $$1 \text{ km/h} = \frac{1000 \text{ [m]}}{3600 \text{ [s]}} = 0.278 \text{ m/s.}$$

Hence $$100 \text{ km/h} = 0.278 \times 100 = 27.8 \text{ m/s.}$$

Example 6.2 *If an aeroplane travels a distance of* 8000 *km at a constant speed in* 12 *h, calculate (a) its speed in metres/second, (b) the number of kilometres travelled in* 20 *min and (c) the time taken to travel* 100 *km.*

(a) $$\text{Distance travelled} = 8000 \times 1000 = 8 \times 10^6 \text{ m}$$
$$\text{and time taken} = 12 \times 3600 = 43\,200 \text{ s,}$$

$\therefore$ $$\text{speed} = \frac{8 \times 10^6 \text{ [m]}}{43\,200 \text{ [s]}} = 185 \text{ m/s.}$$

(b) $$\text{Since speed} = \frac{8000 \text{ [km]}}{12 \text{ [h]}} = 667 \text{ km/h}$$

$\therefore$ distance travelled in 20 min $= 667 \text{ [km/h]} \times (20/60) \text{ [h]}$
$$= 222 \text{ km.}$$

Alternatively, since speed $= 185 \text{ m/s}$
$\therefore$ distance travelled in 20 min $= 185 \text{ [m/s]} \times (20 \times 60) \text{ [s]}$
$$= 222\,000 \text{ m}$$
$$= 222 \text{ km.}$$

(c) From expression (6.1),

$$\text{time to travel } 100 \text{ km} = \frac{100 \text{ [km]}}{667 \text{ [km/h]}} = 0.15 \text{ h}$$
$$= 0.15 \times 60 = 9 \text{ min.}$$

It will be noted that we have not taken any account of direction of motion of the aeroplane—in other words, the speed of a body is independent of the direction of motion.

6.3 Graphs relating distance, time and speed

The relationships between distance and time and between speed and time can usefully be represented by simple graphs.

Distance/time and speed/time graphs for constant speed. Fig. 6.2(a) is a graph showing the distance travelled during a period of 20 s by a body moving at a constant speed. The *slope* of this straight-line graph, calculated in terms of the units used for the two axes, is 100 m divided by 20 s, i.e. 5 m/s, and therefore represents the *speed* at which the body is moving.

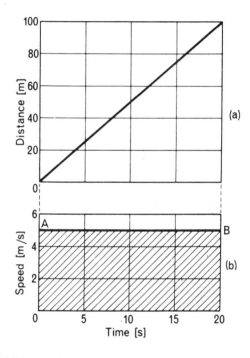

Fig. 6.2 Distance/time and speed/time graphs for constant speed.

The horizontal line AB in fig. 6.2(b) is a graph representing the constant speed of 5 m/s derived from fig. 6.2(a)

The area shown shaded in fig. 6.2(b)

$$= \text{(speed in metres/second)} \times \text{(time in seconds)}$$
$$= 5\ [\text{m/s}] \times 20\ [\text{s}] = 100\ \text{m}$$
$$= \text{distance travelled.}$$

Distance/time and speed/time graphs for varying speed. Fig. 6.3(a) is a distance/time graph for a body which travels a distance of 100 m in 20 s at varying speeds. The *average* speed is 100 [m]/20 [s] = 5 m/s. The initial and the final speeds are zero; consequently the slopes of the graph at the beginning and the end of the period must

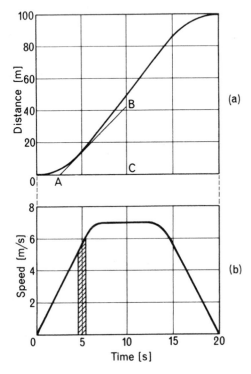

Fig. 6.3 Distance/time and speed/time graphs for varying speed.

be zero. The slope at any intermediate instant is obtained by drawing a tangent to the graph at that instant; for example, at 5 seconds, the tangent is shown by line AB.

$$\text{Slope of AB} = \frac{\text{BC}}{\text{AC}} = \frac{42 \ [\text{m}]}{(10 - 2\cdot5) \ [\text{s}]} = 5\cdot6 \ \text{m/s}.$$

Similarly, it is found that the slope of the graph between 7 s and 13 s is constant at about 7·1 m/s. Thus, by drawing tangents at various points of the distance/time graph of fig. 6.3(a), the speed/

time graph of fig. 6.3(b) can be derived.

During the 1 second from 4·5 s to 5·5 s, the average speed is practically 5·6 m/s, so that the distance travelled during that 1 second = 5·6 [m/s] × 1 [s] = 5·6 m, and is represented by the area of the shaded strip in fig. 6.3(b). Similarly for all the thin strips into which we might divide the area under the graph. Hence the total area enclosed by the speed/time graph of fig. 6.3(b) represents the total distance travelled.

If the only information available about the movement of a body was the speed/time graph such as that shown in fig. 6.4, the simplest method of determining the average speed is by means of mid-ordinates. Thus, if the base line is divided into, say, 6 equal lengths, as in fig. 6.4, and the mid-ordinates v_1, v_2, etc., are drawn and measured, then:

$$\text{average speed} = v = \frac{v_1 + v_2 + v_3 + v_4 + v_5 + v_6}{6}$$

If t be the length of the base of the speed/time graph,

$$\text{distance travelled} = \text{area enclosed by graph}$$
$$= \text{average speed} \times \text{time}$$
$$= vt.$$

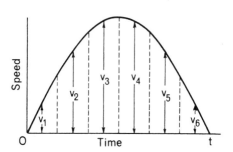

Fig. 6.4 Graphical determination of the average speed.

The larger the number of ordinates used, the more accurate the result.

6.4 Linear velocity

It was pointed out in section 6.2 that the speed of a body can be stated without any reference to the direction of movement of that body. Consequently, *speed* is a *scalar* quantity. If, however, we specify the *direction* of motion as well as the *speed* of the body, the quantity is then termed the *velocity* of the body; thus, if a car is travelling in a northerly direction at a speed of 40 kilometres/hour, the velocity is said to be 40 km/h northwards. Since velocity has both magnitude and direction, it is a *vector* quantity and can be represented by a straight line drawn to scale in the direction of the velocity. If a body travels a distance s in a constant direction in time t, and if v is the average velocity, then:

$$v = s/t \tag{6.1}$$

and the speed and velocity are numerically the same.

6.5 Acceleration

When the velocity of a body is increasing, the body is said to be accelerating, whereas if the velocity is decreasing, the body is said to be retarding. Retardation may be regarded as negative acceleration. Suppose that the velocity of a train on a straight horizontal track increases by 1·5 m/s every second from standstill until the train attains a speed of 30 m/s, then at the end of 1 second the speed is 1·5 m/s, at the end of the next second it is 3 m/s, at the end of the third second it is 4·5 m/s, etc., until at the end of 20 s the velocity is 30 m/s, and the variation of velocity with time can be represented by the straight line OA in fig. 6.5. It follows that *acceleration* can be defined as *the rate of change of velocity*; and

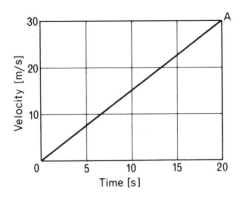

Fig. 6.5 Uniform acceleration.

when the rate of change remains constant, as in fig. 6.5, the acceleration is said to be *uniform*.

Suppose the initial velocity of a body moving in a straight line to be u, as shown in fig. 6.6, and suppose the velocity to increase at a uniform rate to v in time t; then, if the acceleration is represented by the symbol a,

$$\text{change of velocity} = v - u$$

and $\quad\quad\quad$ acceleration $= a =$ rate of change of velocity

$$= \frac{\text{change of velocity}}{\text{time}}$$

$$= \frac{v - u}{t}$$

∴ $\quad\quad\quad\quad\quad\quad v = u + at \quad\quad\quad\quad\quad (6.2)$

If u and v are expressed in metres/second and t is in seconds, then the acceleration is in metres per second every second, i.e. metres per second squared, the symbol being m/s^2 (not m/s/s).

Since the velocity is assumed to vary at a uniform rate between u and v, it is represented by the straight line AB in fig. 6.6. The average velocity is the mean of u and v, namely $\frac{1}{2}(u + v)$; hence, if s represents the distance travelled, we have:

$$s = \text{average velocity} \times \text{time}$$
$$= \tfrac{1}{2}(u + v)t$$

Substituting the value for v from equation (6.2), we have:

$$s = \tfrac{1}{2}(u + u + at)t$$
$$= ut + \tfrac{1}{2}at^2 \quad\quad\quad\quad (6.3)$$

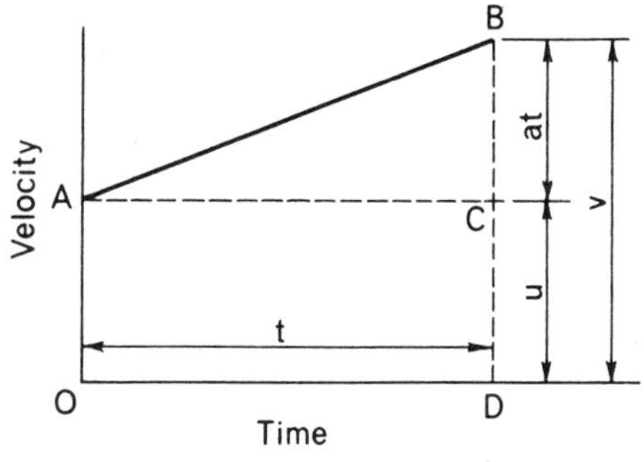

Fig. 6.6 Uniform acceleration.

We can derive an expression for v in terms of u, a and s by squaring the two sides of equation (6.2) and then substituting the value of s from equation (6.3), thus:

$$
\begin{aligned}
v^2 &= (u + at)^2 \\
&= u^2 + 2uat + (at)^2 \\
&= u^2 + 2a(ut + \tfrac{1}{2}at^2) \\
&= u^2 + 2as
\end{aligned}
\tag{6.4}
$$

It should be noted that in the above expressions, a is positive when the body is accelerating and negative when it is retarding.

6.6. Acceleration of a falling body

When a body falls, the force of attraction between the body and the earth causes its velocity to increase. The pull with which a body is attracted towards the earth's centre is referred to as *gravitational force*. It is well known that a feather falls less rapidly than a piece of metal, but this difference is due to the relatively large surface of the feather so that the air resistance is comparatively large. It was Sir Isaac Newton who first proved experimentally that all bodies fall with the same acceleration so long as their movement is not impeded by any resistance. He dropped a feather and a coin in a long vertical glass tube from which practically all the air had been extracted and showed that they fell at the same rate.

The acceleration of a freely falling body is referred to as the acceleration due to gravity and is represented by the symbol g. It has already been stated in section 1.8 that owing to the radius of the earth being slightly smaller at the north and south poles than it is at the equator, the gravitational pull is slightly higher at the poles than it is at the equator. Consequently, the value of g is about $9 \cdot 832$ m/s² at the poles and about $9 \cdot 780$ m/s² at the equator. In London, at sea level, the value of g is almost exactly $9 \cdot 81$ m/s².

It follows from equation (6.2) that if a body, *initially at rest*, falls freely for time t, the final velocity v is given by

$$
v = gt
\tag{6.5}
$$

If s is the distance travelled, then from equation (6.3):

$$
\begin{aligned}
s &= \text{average velocity} \times \text{time} \\
&= \tfrac{1}{2}vt = \tfrac{1}{2}gt^2
\end{aligned}
\tag{6.6}
$$

and from equation (6.4):

$$v = \sqrt{(2gs)} \qquad (6.7)$$

Example 6.3 *A train has a uniform acceleration of* 0.2 *m/s² along a straight track. Calculate* (a) *the velocity after an interval of* 16 *s from standstill,* (b) *the time required to attain a velocity of 50 km/h,* (c) *the distance travelled from standstill until the train attains a velocity of 50 km/h and* (d) *the time taken for the velocity to increase from 30 km/h to 50 km/h and the distance travelled during that time.*

(a)   Since $v = u + at$ (6.2)

$\therefore$ $v = 0 + 0.2$ [m/s²] $\times$ 16 [s] $= 3.2$ m/s.

(b)   50 km/h $= 50\ 000$ [m]/3600 [s] $= 13.9$ m/s.

Substituting in expression (6.2) given above, we have:

$$13.9 \text{ [m/s]} = 0 + 0.2 \text{ [m/s}^2\text{]} \times t$$

$\therefore$ $t = 69.5$ s.

(c) Since the acceleration is *uniform,*

$$s = \tfrac{1}{2}(u + v)t$$
$$= \tfrac{1}{2}(0 + 13.9) \text{ [m/s]} \times 69.5 \text{ [s]}$$
$$= 483 \text{ m} = 0.483 \text{ km.}$$

(d)   30 km/h $= 30\ 000$ [m]/3600 [s] $= 8.33$ m/s.

Substituting in expression (6.2) given above, we have:

$$13.9 \text{ [m/s]} = 8.33 \text{ [m/s]} + 0.2 \text{ [m/s}^2\text{]} \times t$$

$\therefore$ $t = 27.85$ s.

Since the acceleration is uniform,

$$s = \tfrac{1}{2}(u + v)t$$
$$= \tfrac{1}{2}(8.33 + 13.9) \text{ [m/s]} \times 27.85 \text{ [s]}$$
$$= 310 \text{ m} = 0.31 \text{ km.}$$

Example 6.4 *A motor car travelling at 50 km/h on dry level surface should be able to stop in 14 metres. Assuming the retardation to be uniform, calculate* (a) *its value and the corresponding braking time and* (b) *the distance travelled during the first second after the application of the brakes.*

(a) 50 km/h = 50 000 [m]/3600 [s] $\eqsim$ 13·9 m/s.

Since $v^2 = u^2 + 2\,as$ (6.4)

$\therefore$ $0 = (13\cdot9)^2\ [\text{m/s}]^2 + 2a \times 14\ [\text{m}]$

so that $a = -\ 6\cdot9\ \text{m/s}^2$,

i.e. the retardation is 6·9 m/s².

Since $v = u + at$ (6.2)

$\therefore$ $0 = 13\cdot9\ [\text{m/s}] + (-\ 6\cdot9)\ [\text{m/s}^2] \times t$

so that $t = 2\cdot02$ s.

Alternatively, since the retardation is uniform,

$$s = \tfrac{1}{2}\,(u + v)t$$

$\therefore$ $14\ [\text{m}] = \tfrac{1}{2}\,(13\cdot9 + 0)\ [\text{m/s}] \times t$

so that $t = 2\cdot02$ s.

(b) Since $s = ut + \tfrac{1}{2}at^2$ (6.3)

$\therefore$ $s = 13\cdot9\ [\text{m/s}] \times 1\ [\text{s}] + \tfrac{1}{2} \times (-\ 6\cdot9)\ [\text{m/s}^2] \times 1^2\ [\text{s}]^2$

$= 10\cdot45$ m.

Example 6.5 *A stone is dropped from a tower,* 100 *m high. Assuming* g = 9·81 *m/s² and the air resistance to be negligible, calculate* (a) *the time taken to reach the ground,* (b) *the velocity of the stone when it hits the ground and* (c) *the distance through which the stone falls during the first* 2 *seconds.*

(a) Since $s = \tfrac{1}{2}gt^2$ (6.6)

$\therefore$ $100\ [\text{m}] = \tfrac{1}{2} \times 9\cdot81\ [\text{m/s}^2] \times t^2$

$\therefore$ $t^2 = 20\cdot4\ \text{s}^2$

and $t = 4\cdot52$ s.

(b) Since $v = gt$ (6.5)

$\therefore$ $v = 9\cdot81\ [\text{m/s}^2] \times 4\cdot52\ [\text{s}]$

$= 44\cdot3$ m/s.

(c) Substituting in expression (6.6) given above, we have:

$$s = \tfrac{1}{2} \times 9\cdot81\ [\text{m/s}^2] \times (2)^2\ [\text{s}]^2$$
$$= 19\cdot62 \text{ m}.$$

Example 6.6 *A cricket ball is thrown vertically upwards at a velocity of* 20 *m/s. Calculate* (a) *the time taken to reach the maximum height and* (b) *the maximum height attained. Assume* g = 9·81 *m/s² and the air resistance to be negligible.*

(*a*) While the ball is moving upwards, the retardation is 9.81 m/s^2, i.e. the acceleration is -9.81 m/s^2.

Since $\qquad\qquad v = u + at$ (6.2)

$\therefore \qquad\qquad 0 = 20 \,[\text{m/s}] + (-9.81) \,[\text{m/s}^2] \times t$

so that $\qquad\quad t = 2.04$ s.

(*b*) Since the retardation is uniform,

$$s = \tfrac{1}{2}(u + v)t$$
$$= \tfrac{1}{2}(20 + 0)\,[\text{m/s}] \times 2.04\,[\text{s}] = 20.4 \text{ m}.$$

6.7 Relative velocity

When we speak of the velocity of a body, we generally mean its velocity relative to the earth, which is itself moving at a high speed; i.e. we think of the rate of displacement as if the earth were at rest. Similarly we sometimes speak of the velocity of one body relative to another body which may be moving at a known velocity relative to the earth. For simplicity, we shall limit our discussion of relative velocity to velocities in, or parallel to, a single straight line.

If a train A is travelling, say, east at 80 km/h, then relative to a second train B travelling east at 50 km/h on a parallel track, the first train is moving at:

$$80 - 50 = 30 \text{ km/h}.$$

To an observer in the second train, A would *appear* to be travelling at 30 km/h eastward; while to an observer on A, train B would appear to be travelling westward at 30 km/h.

If the second train B had been travelling *westward* at 50 km/h, the first train A would be travelling eastward *relative* to the second train at:

$$80 - (-50) = 130 \text{ km/h},$$

and would *appear* to be travelling eastward at 130 km/h to an observer in the second train B. To an observer in A, train B would appear to be travelling westward at 130 km/h.

6.8 Resultant of two velocities

Suppose an aeroplane to be flying at a constant velocity such that in *perfectly still air* it would travel northward at 500 km/h relative to the ground, and suppose the air to have a constant velocity of

80 km/h in an easterly direction relative to the ground. Under such circumstances, the plane will still be travelling at 500 km/h northward relative to the air.

We can represent these two velocities vectorially as in fig. 6.7, where vector OA represents the air velocity of 80 km/h easterly relative to the *ground*, and vector AB represents the plane velocity of 500 km/h northerly relative to the *air*. The vector sum of OA and AB is OB, i.e. OB represents the velocity of the aeroplane relative to the *ground*.

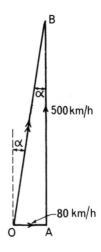

Fig. 6.7 Resultant of two velocities.

It will be noted that we started from point O with the vector OA representing a velocity component relative to the *ground*. Consequently the vector OB gives the resultant velocity also relative to the *ground*, i.e. relative to the object we regard as stationary.

From fig. 6.7 it will be seen that:

$$OB = \sqrt{(OA^2 + AB^2)}$$
$$= \sqrt{(80^2 + 500^2)} = 506 \text{ km/h.}$$

If α is the angle between OB and AB,

$$\tan \alpha = OA/AB = 80/500 = 0.16,$$
$$\therefore \qquad \alpha = 9.1°.$$

Hence the resultant velocity of the aeroplane is 506 km/h, in a direction 9·1° east of north.

Let us now consider the general case of a body possessing two simultaneous velocities represented by vectors OA and AB in fig. 6.8, and suppose θ to be the angle between the directions of the two velocities. The resultant velocity is the vector sum of OA and AB, namely OB in fig. 6.8. Double arrowheads have been inserted on OB to indicate that it represents the resultant of *two* velocities.

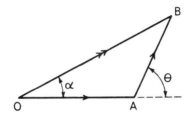

Fig. 6.8 Resultant of two velocities.

If both the magnitude and the direction of the resultant velocity is required, the simplest method at this stage is to draw the diagram to a large scale and measure the length of OB and the value of angle α. If, however, only the magnitude of the resultant is required, this can be calculated from the relationship:

$$OB = \sqrt{(OA^2 + AB^2 + 2 \times OA \times AB \cos \theta)}$$

Example 6.7 *A river is flowing at 1 m/s and is 100 m wide. If a man rows a boat at 2 m/s in still water, determine the direction in which he must row at the same pace in order to reach a point D on the other bank exactly opposite the starting point C (fig. 6.9). Also, calculate the time taken to cross the river.*

If vector OA in fig. 6.9 represents the velocity of the river, then in order that the boat may travel from point C on one bank to point D on the opposite bank, the vector representing the resultant velocity must lie along line CD.

A graphical solution can be obtained by using a scale of, say, 1 mm to 0·05 m/s. Thus OA is drawn 20 mm long to represent the velocity of the river relative to its banks. With point A as centre, draw an arc having a radius of 40 mm, cutting CD at B, to represent the speed of the boat in still water. Then line AB represents the velocity of the boat relative to the *river* and OB represents the resultant velocity of the boat relative to the *river banks*. From the diagram it is found that angle θ is 30°, i.e. the man must row the boat in a direction making an angle of 30° with CD.

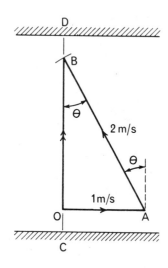

Fig. 6.9 Example 6.7.

Alternatively, angle θ can be calculated from the fact that angle AOB is a right angle; thus:

$$\sin \theta = OA/AB = 1 \, [\text{m/s}]/2 \, [\text{m/s}] = 0\cdot5$$
$$\therefore \qquad \theta = 30°.$$

The resultant velocity is represented by vector OB and its value can be determined graphically or calculated thus:

$$OB = AB \cos 30°$$
$$= 2 \, [\text{m/s}] \times 0\cdot866 = 1\cdot732 \, \text{m/s}.$$

Hence,   time taken to $\Big\}$ = $\dfrac{\text{width of river}}{\text{velocity in direction CD}}$
           cross river

$$= \frac{100 \, [\text{m}]}{1\cdot732 \, [\text{m/s}]} = 57\cdot7 \, \text{s}.$$

6.9 Resolution of a velocity into two component velocities

In the preceding section it was shown how the resultant of two simultaneous velocities in different directions can be determined. Conversely, a velocity can be resolved into two components, the only condition being that the resultant of the two vectors representing the component velocities must be the same in magnitude and direction as the vector representing the original velocity. The procedure is similar to the resolution of a force discussed in section 2.9.

In practice, the *directions* of the components are generally specified; and these directions are usually at right angles to each other, in which case they are referred to as *rectangular* components. Thus, if vector OB in fig. 6.10 represents the magnitude and direction of

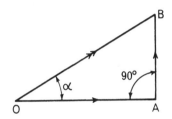

Fig. 6.10 Resolution of velocity into rectangular components.

the velocity with which, say, a cricket ball is thrown into the air, this velocity can be resolved into a horizontal component OA and a vertical component AB. If α be the angle between OB and the horizontal component,

$$\left.\begin{array}{l}\text{horizontal component}\\\text{of the velocity}\end{array}\right\} = \text{OA} = \text{OB} \cos \alpha \qquad (6.8)$$

and
$$\left.\begin{array}{l}\text{vertical component}\\\text{of the velocity}\end{array}\right\} = \text{AB} = \text{OB} \sin \alpha \qquad (6.9)$$

Example 6.8 *If a ball is thrown with a velocity of 20 m/s in a direction inclined at 60° to the horizontal, determine* (a) *the horizontal and vertical components of the velocity,* (b) *the time taken for the ball to return to the ground and* (c) *the horizontal distance travelled before the ball touches the ground. Assume the ground to be level, g to be 9·81 m/s² and the air resistance to be negligible.*

(*a*) From expressions (6.8) and (6.9),

horizontal component of velocity = 20 cos 60° = 10 m/s
and vertical component of velocity = 20 sin 60° = 17·32 m/s.

(*b*) Let us consider the vertical component of the velocity.

Since $$v = u + at \tag{6.2}$$
and $a = -9\cdot81$ m/s²,
$$0 = 17\cdot32 \text{ [m/s]} + (-9\cdot81) \text{ [m/s}^2\text{]} \times t$$

so that $\quad$ time taken to reach $\left.\begin{array}{}\\ \\\end{array}\right\} = t = \dfrac{17\cdot32 \text{ [m/s]}}{9\cdot81 \text{ [m/s}^2\text{]}} = 1\cdot77$ s,
$\quad$ maximum height

∴ $\quad$ total time for ball to $\left.\begin{array}{}\\ \\\end{array}\right\} = 2 \times 1\cdot77 = 3\cdot54$ s.
$\quad$ return to ground

(*c*) Since the ball has a horizontal velocity of 10 m/s and remains in the air for 3·54 s,

horizontal distance travelled = 10 × 3·54 = 35·4 m.

This example illustrates the usefulness of resolving a velocity into rectangular components.

Summary of Chapter 6

The slope of a distance/time graph at any instant represents the speed at that instant, and the area enclosed by a speed/time graph represents the distance travelled.

Acceleration is the rate of change of velocity and the value of the acceleration due to gravity at sea level in the vicinity of London is almost exactly 9·81 m/s². If the velocity of a body is accelerated uniformly from u to v in time t, and if s is the distance travelled and a is the acceleration,

$$\text{average velocity} = \tfrac{1}{2}(u + v)$$
$$v = u + at \tag{6.2}$$
$$s = \tfrac{1}{2}(u + v)t$$
$$= ut + \tfrac{1}{2}at^2 \tag{6.3}$$
and $$v^2 = u^2 + 2as \tag{6.4}$$

If the two simultaneous velocities of a body are represented in magnitude and direction by two vectors, the resultant velocity is represented in magnitude and direction by the vector sum of the two vectors.

A velocity can be resolved into two component velocities. If the components are at right angles to each other, they are termed rectangular components.

EXAMPLES 6

1. A car travels 20 km southwards and then travels another 30 km westwards. What is the displacement of the car from its initial position?
2. A person walks a distance of 8 km in a direction 20° east of north, and then walks another 6 km in a direction 50° south of east. Determine graphically the value and direction of the final displacement relative to the starting point.
3. An aeroplane flies 160 km in a southerly direction. Its direction is then changed to SE. With the aid of a vector diagram drawn to scale, determine the distance travelled in the new direction when its resultant displacement is 260 km.
4. A boat is rowed a distance of 2 km in a direction 40° north of east. Its direction is then changed to 20° west of south. Determine graphically the distance it has to travel in that direction for the final displacement to be 30° south of east. What is the value of that final displacement?
5. Convert the following speeds into metres/second: (a) 40 km/h, (b) 5000 mm/min, (c) 15 km/min.
6. Convert the following speeds into kilometres/hour: (a) 20 m/s, (b) 90 mm/s, (c) 3 km/s, (d) 15 m/min.
7. If a train is travelling at 90 km/h, what is its speed in (a) kilometres/minute, (b) metres/second?
8. If a man is walking at 1·8 m/s, calculate the distance covered in 2 hours.
9. If a person runs a distance of 0·7 km in 3 min, what is his average speed in (a) kilometres/hour, (b) metres/second?
10. If a car is travelling at 25 m/s, calculate (a) the speed in kilometres/hour and (b) the distance, in kilometres, covered in 12 min.
11. A cage is lowered down a pit shaft, 600 m deep, in 46 s. What is its average speed in (a) metres/second, (b) kilometres/hour?
12. The minute hand of Big Ben is 4·27 m long. Calculate the speed of the tip of this hand in (a) metres/minute, (b) millimetres/second.
13. A car travels at 50 km/h for the first half hour and at 80 km/h for the following hour. Calculate (a) the total distance travelled and (b) the average speed.
14. A train travels the first 80 km of its journey at an average speed of 60 km/h. What must be its average speed over the remaining 50 km in order that the average speed over the whole journey may be 70 km/h?
15. A cyclist covers the first 24 km of his journey in 80 min. After a rest of 10 min, he covers the remaining 20 km in 70 min. What is his average speed, in kilometres/hour, for the whole journey?

16. A train, starting from rest, covers the following distances x metres in times t seconds:

t	0	5	11	18	22	27	31	38	46	50
x	0	3	16	52	79	115	137	158	168	170

Plot the distance/time graph and determine the approximate speeds in metres/second, after 5, 15, 25, 35 and 45 seconds from the start. Using this data, plot the speed/time graph for the whole period.

What is the average speed, in metres/second, over the 50 s?

17. A car, starting from rest, accelerates uniformly at 2 m/s² for 10 s. Calculate (*a*) the distance travelled, in metres, and (*b*) the final speed, in metres/second.

18. The speed of a motor vehicle falls uniformly from 100 km/h to zero at 3 m/s². Calculate (*a*) the time taken for the vehicle to stop, (*b*) the distance travelled.

19. A train, starting from rest, reaches a speed of 18 m/s in 2 min. Assuming the acceleration to be uniform, calculate (*a*) the value of the acceleration in metres per second squared and (*b*) the distance travelled, in kilometres.

20. The velocity of a body increases uniformly from 15 km/h to 40 km/h while it travels 100 m. Calculate (*a*) the acceleration, in metres per second squared and (*b*) the time taken, in seconds.

21. A stone is dropped down a shaft, 160 m deep. Calculate (*a*) the time taken for the stone to reach the bottom and (*b*) the velocity of the stone as it reaches the bottom. Assume $g = 9 \cdot 81$ m/s².

22. A cricket ball, thrown vertically upwards, returns to the ground in 4 s. Calculate (*a*) the height, in metres, reached by the ball and (*b*) the velocity, in metres per second, with which it is thrown.

23. A ball is thrown vertically upwards at 40 m/s. Calculate (*a*) the greatest height attained, (*b*) the time taken to return to the ground, (*c*) its velocity after 3 s and after 6 s and (*d*) the distance travelled during the first 3 s.

24. The following table shows how the speed of a car varied with time:

Speed (km/h)	0	19	35	48	48	27	0
Time (s)	0	3	6	9	12	15	18

Assuming these speed values to be joined by straight lines, plot a graph of speed against time and estimate the distance travelled. Also determine the values of the acceleration during the first 3 s and of the retardation during the last 3 s, in metres per second squared. What is the average speed in kilometres per hour?

25. A train makes a journey between two stations in 8 min. The speed of the train at 1-minute intervals is:

Time (min)	0	1	2	3	4	5	6	7	8
Speed (km/h)	0	18	34	45	48	48	42	24	0

Draw the speed/time diagram and determine from the diagram (*a*) the approximate value of the acceleration 2 min after starting and (*b*) the distance between the stations.

26. From a point A, a body travels for 2 s at a constant velocity of 12 m/s due north. It then travels for 5 s due east whilst its velocity decreases uniformly from 12 m/s to 6 m/s. What is the resultant displacement from A? Also, determine the average velocity, in magnitude and direction, of the body for it to return along a straight path to point A in 3 s.
27. A railway carriage is travelling along a straight track at 12 m/s and a ball is rolled at 4 m/s across the floor of the carriage at right angles to the direction of motion of the train. Calculate the magnitude of the resultant velocity of the ball and its direction relative to that of the carriage.
28. A boat is rowed with a velocity of 7 km/h at right angles to a river flowing at 3 km/h. The river is 60 m wide. How far down the river will the boat reach the opposite bank below the point at which it was originally directed and what is the actual velocity of the boat?
29. A boat is rowed on a river so that its speed in still water would be 3 m/s. If the river flows at the rate of 2 m/s, determine at what inclination to the direction of flow of the river must the boat be headed so that the motion of the boat may be at right angles to the current. If the width of the river is 40 m, how long will it take for the boat to cross from one bank to the other?
30. A body is travelling in a straight line at 15 m/s. Determine the rectangular components of the velocity such that one component is inclined at an angle of 30° to the direction of motion.
31. A shell is fired with a velocity of 600 m/s in a direction inclined at 70° to the horizontal. Calculate (*a*) the horizontal and vertical components of the velocity, (*b*) the greatest height above ground attained by the shell and (*c*) the horizontal distance travelled by the shell before it touches ground. Assume the ground to be level and the air resistance to be negligible.
32. A body is travelling in a straight line at 40 m/s. Determine graphically or otherwise the component velocities such that one component is making an angle of 20° with the direction of motion and the angle between the two components is 70°.
33. Two trains are travelling on parallel tracks at speeds of 30 km/s and 80 km/s respectively. Calculate the relative speed of the trains if they are travelling (*a*) in the same direction and (*b*) in opposite directions.

ANSWERS TO EXAMPLES 6

1. 36·1 km, 33·7° south of west.
2. 7·21 km, 23·9° north of east.
3. 121 km.
4. 1·9 km, 1·02 km.
5. 11·1 m/s, 0·0833 m/s, 250 m/s.
6. 72 km/h, 0·324 km/h, 10 800 km/h, 0·9 km/h.
7. 1·5 km/min, 25 m/s.
8. 12·96 km.
9. 14 km/h, 3·89 m/s.
19. 0·15 m/s², 1·08 km.
20. 0·53 m/s², 13·1 s.
21. 5·71 s, 56 m/s.
22. 19·62 m, 19·62 m/s.
23. 81·5 m; 8·15 s; 10·57 m/s upward, 18·9 m/s downward; 75·8 m.
24. 147·5 m, 1·76 m/s², 2·5 m/s², 29·5 km/h.
25. 0·068 m/s² (approx.), 4·4 km (approx.).

10. 90 km/h, 18 km.
11. 13·04 m/s, 47 km/h.
12. 0·447 m/min, 7·45 mm/s.
13. 105 km, 70 km/h.
14. 95·5 km/h.
15. 16·5 km/h.
16. 1·1 m/s, 5·3 m/s, 7·0 m/s,
    2·8 m/s, 0·8 m/s; 3·4 m/s.
17. 100 m, 20 m/s.
18. 9·3 s, 129 m.

26. 51 m, 17 m/s, 61·9° west of
    south.
27. 12·65 m/s, 18·4°.
28. 25·7 m, 7·62 km/h.
29. 131·8°, 17·9 s.
30. 13 m/s, 7·5 m/s.
31. 205 m/s, 564 m/s; 16·2 km;
    23·6 km.
32. 32·6 m/s, 14·55 m/s.
33. 50 km/h, 110 km/h.

CHAPTER 7

Newton's laws of motion

7.1 First law of motion

A body continues in its state of rest or of uniform motion in a straight line unless it is compelled by an external force to change that state.

This law may be called the 'law of inertia'. A body will not change its state of rest or of motion in a straight line unless compelled to do so; i.e. it resists any change of velocity in magnitude or direction. From this law, we may define a *force* as *any push or pull which changes or tends to change the state of rest of a body or its uniform motion in a straight line.*

7.2 Momentum

This is the name given to the product of the mass m of a body and its velocity v,

i.e. $\qquad$ momentum $= mv$.

Mass has no direction and is therefore a scalar quantity; but velocity has magnitude and direction and is therefore a vector quantity. Hence momentum must also be a vector quantity; and a vector which represents the velocity of a body can also, to a different scale, represent the momentum of that body. Values of momentum can therefore be added and resolved vectorially in the same way as for velocity (sections 6.8 and 6.9).

7.3 Second law of motion

When a body is acted upon by an external force, the rate of change of momentum is proportional to the force and takes place in the direction of the force.

This law is in effect the definition of a force in respect to its effect on a body.

If a force F acts upon a body of mass m for a time t and causes its velocity in the direction of the force to increase from v_1 to v_2,

$$\text{initial momentum} = mv_1$$

and $$\text{final momentum} = mv_2,$$

∴ $$\left.\begin{array}{c}\text{average rate of change}\\\text{of momentum}\end{array}\right\} = (mv_2 - mv_1)/t$$

$$= m\,(v_2 - v_1)/t = ma,$$

where $a = $ average acceleration
                                during time t.

Hence, from Newton's Second Law,

$$F \propto ma$$
$$= ma \times \text{a constant} \qquad (7.1)$$

It has already been stated in section 1.8 that the SI unit of force is the *newton*. This has been defined as *the force required to give a mass of* 1 *kg an acceleration of* 1 *m/s²* in order to give a value of one for the constant. Substituting $m = 1\,\text{kg}$, $a = 1\,\text{m/s}^2$ and $F = 1\,\text{N}$ in expression (7.1), we have:

$$1\;[\text{N}] = 1\;[\text{kg}] \times 1\;[\text{m/s}^2] \times \text{a constant}.$$

Hence the constant is unity.

It follows that if F be the force, in newtons, required to give a mass m, in kilograms, an acceleration a, in metres per second squared, then:

$$F = ma \qquad (7.2)$$

Example 7.1 *A force of* 50 *N is applied to a mass of* 200 *kg. Calculate the acceleration.*

Substituting for F and m in expression (7.2), we have:

$$50\;[\text{N}] = 200\;[\text{kg}] \times a$$
∴ $$a = 0{\cdot}25\;\text{m/s}^2.$$

Example 7.2 *A mass of* 3 *Mg is to be given an acceleration of* 2 *m/s². Calculate the force required.*

$$\text{Mass} = 3\;\text{Mg} = 3000\;\text{kg}$$
∴ $$F = 3000\;[\text{kg}] \times 2\;[\text{m/s}^2]$$
$$= 6000\;\text{N} = 6\;\text{kN}.$$

Example 7.3 *A Diesel engine pulling a train along a level track has its oil supply cut off when the train is travelling at* 60 *km/h. It is observed that the speed falls to* 40 *km/h after the train has travelled a distance of* 1200 *m. The mass of the engine and carriages is* 80 *Mg. Assuming the retardation to be uniform, calculate the total force resisting motion.*

$$60 \text{ km/h} = 60\ 000 \text{ [m]}/3600 \text{ [s]} = 16 \cdot 67 \text{ m/s}$$

and
$$40 \text{ km/h} = 40\ 000 \text{ [m]}/3600 \text{ [s]} = 11 \cdot 11 \text{ m/s}.$$

Since
$$v^2 = u^2 + 2\ as \qquad (6.4)$$

∴
$$(11 \cdot 11)^2 \text{ [m/s]}^2 = (16 \cdot 67)^2 \text{ [m/s]}^2 + 2a \times 1200 \text{ [m]}$$

so that
$$a = -0 \cdot 0642 \text{ m/s}^2.$$

$$\text{Mass of train} = 80 \text{ Mg} = 80\ 000 \text{ kg}.$$

Since
$$F = ma$$

∴
$$\textit{retarding} \text{ force} = 80\ 000 \text{ [kg]} \times 0 \cdot 0642 \text{ [m/s}^2\text{]}$$
$$= 5136 \text{ N} = 5 \cdot 136 \text{ kN}.$$

7.4 Third law of motion

To every force there is an equal and opposite force. This law may be stated in another way. If any body A exerts a certain force on another body B, then B exerts on A a force of equal magnitude but in the opposite direction. This is equally true if A and B are two parts of the same body, as already discussed in section 2.1.

This law applies to bodies whether they are at rest or in motion. Thus, if a beam or other body exerts a certain downward force upon a support, the support exerts an equal upward force on the beam—the latter force being often referred to as the reaction of the support (section 3.5).

The fact that this relationship is equally true in regard to two bodies in motion is not so widely realized, e.g. that the backward pull of a trailer on a motor vehicle is equal to the forward pull of the vehicle on the trailer. Let us look a little more closely at the state of affairs when a motor tractor is pulling a trailer along a level road. If the tractor is moving at a constant velocity, the forward pull exerted by the tractor exactly balances the backward pull of the trailer due to friction and wind resistance.

If, however, the forward pull exerted by the tractor exceeds the backward pull due to friction and wind resistance, acceleration occurs. In this case, the forward pull exerted by the tractor may be regarded as being the sum of the following two components: (*a*) the pull F_1 required to haul the trailer at a constant velocity, (*b*) the pull F_2 ($= ma$) required to give the mass m of the trailer an acceleration a. Hence the total *forward* pull exerted *by the tractor on the trailer* is ($F_1 + F_2$) and is exactly equal to the total *backward* pull exerted *by the trailer on the tractor.*

Another example of the application of Newton's Third Law is the recoil of a gun firing a bullet or a shell. The force responsible for the recoil of the gun is exactly equal to that acting on the bullet or shell.

Example 7.4 *A rope supports a mass of 50 kg. Calculate the pull on the rope when the mass is being* (a) *raised,* (b) *lowered, with an acceleration of* 0·5 m/s².

When the mass is *stationary* or moving at a *uniform* velocity,

$$\text{pull on rope} = \text{weight of the mass}$$
$$\simeq 50 \times 9·81 = 490·5 \text{ N}.$$

(*a*) If F be the upward pull, in newtons, when the mass is being raised with an acceleration of 0·5 m/s², the force available for upward acceleration is ($F - 490·5$) newtons. Hence,

$$(F - 490·5) \text{ [N]} = 50 \text{ [kg]} \times 0·5 \text{ [m/s}^2\text{]}$$
so that $\quad\quad\quad\quad\quad F = 515·5 \text{ N}.$

(*b*) If F be the upward pull, in newtons, when the mass is being lowered with an acceleration of 0·5 m/s², the force available for downward acceleration is ($490·5 - F$) newtons. Hence,

$$(490·5 - F) \text{ [N]} = 50 \text{ [kg]} \times 0·5 \text{ [m/s}^2\text{]}$$
so that $\quad\quad\quad\quad\quad F = 465·5 \text{ N}.$

Example 7.5 *What force will a man having a mass of* 70 kg *exert on the floor of a lift* (a) *ascending and* (b) *descending with an acceleration of* 0·6 m/s²?

When the lift is *stationary* or moving at a *uniform* velocity, force exerted by the man on the floor of the lift

$$= \text{weight of the man}$$
$$\simeq 70 \times 9·81 = 686·7 \text{ N}.$$

Hence, by Newton's Third Law,

force exerted by the floor of the lift on the man = 686·7 N.

(*a*) When the lift is ascending with an acceleration of 0·6 m/s²,

force required to accelerate the man upwards
= 70 [kg] × 0·6 [m/s²] = 42 N.

This upward force is provided by an increase in the force exerted by the floor of the lift on the man,

hence total force exerted on the man
= 686·7 + 42 = 728·7 N
= total force exerted by the man on the floor of the lift.

(*b*) When the lift is descending with an acceleration of 0·6 m/s²,

force required to accelerate the man downwards
= 70 [kg] × 0·6 [m/s²] = 42 N.

This downward force is derived from the weight of the man; consequently the net force exerted by the man on the floor of the lift is 42 N less than the value when the lift is stationary or moving at a constant velocity.

Hence, the downward force exerted by the man on the floor of the lift = 686·7 − 42 = 644·7 N.

Summary of Chapter 7

First Law of Motion. A body continues in its state of rest or of uniform motion in a straight line unless it is compelled by an external force to change that state.

Second Law of Motion. When a body is acted upon by an external force, the rate of change of momentum is proportional to the force and takes place in the direction of the force.

Third Law of Motion. To every force there is an equal and opposite force.

Momentum = mass × velocity.

If a force *F*, in newtons, applied to a body of mass *m*, in kilograms, produces an acceleration *a*, in metres/second², then:

$$F = ma \qquad (7.2)$$

EXAMPLES 7

1. Calculate the acceleration produced when a force of 30 N acts on a mass of 50 kg.
2. What is the force required to give a mass of 10 kg an acceleration of 5 m/s²?
3. Calculate the force, in newtons, to give a mass of 500 g an acceleration of 600 mm/s².
4. A body having a mass of 10 kg is travelling in a straight line at 10 m/s. What is its momentum?

   What is the average force required to increase its velocity from zero to 20 m/s in 4 s?
5. A body travelling in a straight line at 20 m/s has a momentum of 300 kg m/s. What is its mass?

   Calculate the force required to reduce the velocity at a uniform rate to 12 m/s in 10 s.
6. A body of mass 20 Mg is acted upon by a force of 15 kN. Assuming the body to be initially at rest, calculate the time taken for the body to acquire a velocity of 500 km/h.
7. A mass of 200 kg is at rest on a perfectly smooth horizontal surface. A constant horizontal force, applied for 5 s, causes it to move 18 m in that time. Calculate (*a*) the value of the force and (*b*) the final speed.
8. A 3-Mg lorry is travelling at 40 km/h on a level road. It is brought to rest with uniform retardation in 10 s. Calculate (*a*) the retarding force in kilonewtons and (*b*) the distance travelled during retardation.
9. Calculate the resistive force required to reduce the speed of a 200-kg mass at a uniform rate from 15 m/s to 5 m/s in 2 min.
10. Calculate the force exerted on the floor of a lift by a person having a mass of 80 kg when the lift is (*a*) ascending and (*b*) descending with an acceleration of 1·5 m/s² in each case.
11. A mass of 25 kg, initially at rest, is lifted 30 m by means of a rope of negligible mass at an acceleration of 2 m/s². Calculate (*a*) the tension in the rope and (*b*) the time taken to lift the load.
12. A block of iron hangs from a spring balance. When the block is stationary, the reading on the balance is 65 N. When the block is hauled vertically upwards with a uniform acceleration, the reading on the balance is 72 N. What is the value of the acceleration?
13. A body has a mass of 1·5 kg. When weighed by means of a spring balance in a moving lift, the balance reading was 13 N. If the lift was accelerating, in what direction was it travelling and what was the value of the acceleration?
14. A pit cage having a mass of 2 Mg is being lowered from rest down a shaft and accelerates uniformly until, when 20 m from the top, the velocity is 400 m/min. Calculate the force in the rope just above the cage. If the cage is then brought to rest uniformly from the speed of 400 m/min in 20 m, what is now the force in the rope?
15. A vehicle is travelling at 100 km/h along a straight level road when the brakes are applied, locking all the wheels. Calculate (*a*) the retardation in

metres per second squared, (*b*) the velocity, in kilometres/hour, after 3 s. Assume the coefficient of friction between the wheels and the road to be constant at 0·6.

16. A railway waggon having a mass of 20 Mg is travelling at 25 km/h along a straight level track. Its wheels are then locked and friction provides a uniform retardation of 1·5 m/s². Calculate (*a*) the distance travelled by the waggon before coming to rest, (*b*) the retarding force due to friction and (*c*) the coefficient of friction between the wheels and the rails.

ANSWERS TO EXAMPLES 7

1. 0·6 m/s²
2. 50 N.
3. 0·3 N.
4. 100 kg m/s, 50 N.
5. 15 kg, 12 N.
6. 185 s.
7. 288 N, 7·2 m/s.
8. 3·33 kN, 55·5 m.
9. 16·67 N.
10. 904·8 N, 664·8 N.
11. 295 N, 5·48 s.
12. 1·055 m/s².
13. 1·14 m/s², downward.
14. 17·4 kN, 21·84 kN.
15. 5·89 m/s², 36·4 km/h.
16. 16·1 m, 30 kN, 0·153.

CHAPTER 8

Angular motion

8.1 Angular displacement

When a shaft rotates, a point on the surface of the shaft moves in a circular path and a line joining the point to the centre of the rotation sweeps out an angle, as in fig. 8.1. This angle is referred to as the angular displacement of the point. The unit of angular displacement is the *radian*, where one radian is the angle subtended at the centre of a circle by an arc equal in length to the radius, as in fig. 8.2.

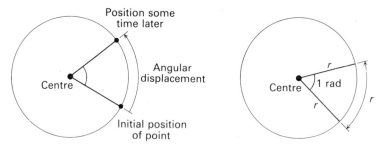

Fig. 8.1 Angular displacement.   Fig. 8.2 The radian.

The relationship between the arc length s, the radius r and the angle subtended θ, is given by

$$s = r\theta \qquad (8.1)$$

Thus when the arc length s equals the radius r we have $s = r = r\theta$ and so θ is 1 radian or 1 rad (the abbreviation for radian). When the arc length is a complete circle, e.g. the point on the rotating shaft having moved through one complete revolution, then as the circumference of a circle is $2\pi r$ we have $s = 2\pi r = r\theta$ and so $\theta = 2\pi$ rad. As one complete revolution is a movement through 360° we must have 2π rad = 360° and hence

$$1 \text{ rad} = \frac{360°}{2\pi}$$

$$1 \text{ rad} = 57·3°$$

Example 8.1 *If a point on a shaft is given an angular displacement of* 1·2 rad, *what is the angular displacement in degrees?*

$$\text{Since } 2\pi \text{ rad} = 360°$$

$$1 \text{ rad} = \frac{360°}{2\pi}$$

Then
$$1·2 \text{ rad} = 1·2 \times \frac{360°}{2\pi}$$

$$= 68·8°$$

Note: it is generally easier to remember that 2π rad equals 360° than 1 rad is 57·3°.

Example 8.2 *Calculate the distance moved by a point on the tyre tread of a car wheel of diameter* 560 mm *if it rotates through* 1·5 rad.

$$\text{Since } s = r\theta$$

and the distance moved by the point is the arc length,

$$\text{distance moved} = r\theta$$
$$= 560 \, [\text{mm}] \times 1·5 \, [\text{rad}]$$
$$= 840 \, \text{mm}$$

Note: The radian unit often does not appear in the final unit, as above, since it is a numerical ratio obtained by dividing an arc, i.e. a distance, by a radius, i.e. another distance and so really does not require a unit.

8.2 Angular velocity

When, for example, a shaft is rotating then a point on the shaft surface has an angular displacement which is varying with time. It can be said to have an angular velocity. *Angular velocity* is defined as the rate of change of angular displacement with time. It is denoted by ω (omega) and has the unit of radian per sec (rad/s).

If a point on a rotating shaft takes a time t to rotate through an angle θ, then the average angular velocity during that time interval is

$$\text{average angular velocity } \omega = \frac{\theta}{t} \tag{8.2}$$

If there is constant angular velocity then equal angular displacements are covered in equal intervals of time—however short the interval.

The angular velocity is related to the *frequency* of rotation. Thus if *n* revolutions are made per second then, since one revolution is an angular displacement of 2π radians, *n* revolutions is an angular displacement of $2\pi n$ radians. This angular displacement occurs in 1 s, hence

$$\text{average angular velocity } \omega = \frac{\text{angular displacement } 2\pi n[\text{rad}]}{\text{time } 1[\text{s}]}$$

$$\omega = 2\pi n \qquad (8.3)$$

Example 8.3 *A flywheel rotates at* 2400 *rev/min. What is its angular velocity?*

$$\text{Since } \omega = 2\pi n$$
and $n = 2400\,\text{rev/min} = 2400/60 = 40\,\text{rev/s}$, then
$$\omega = 2\pi \times 40$$
$$= 251\,\text{rad/s}$$

8.3 Angular acceleration

When the angular velocity of a body is changing there is said to be an angular acceleration. *Angular acceleration* is defined as the rate of change of angular velocity with time. The symbol for angular acceleration is α (alpha) and the unit is radian per second squared (rad/s^2).

If the angular velocity of a rotating object changes from ω_0 to ω_1 in a time interval of t, then the average angular acceleration during that time interval is

$$\text{average acceleration } \alpha = \frac{(\omega_1 - \omega_0)}{t} \qquad (8.4)$$

If there is constant angular acceleration then the angular velocity is changing by equal amounts in equal intervals of time—however short the interval.

Example 8.4 *The angular velocity of a grinding wheel changes from zero to* 150 *rad/s in* 30 *s, what is the average angular acceleration?*

$$\text{Since } \alpha = \frac{(\omega_1 - \omega_0)}{t}$$

$$\alpha = \frac{(150 - 0)}{30\,[s]}\,[rad/s]$$

$$= 5 \cdot 0\,rad/s^2$$

8.4 Relationship between linear and angular motion

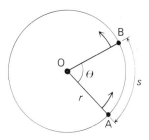

Fig. 8.3

Consider a point on the rim of a wheel of radius r moving with a constant angular velocity ω, as in fig. 8.3. In a time t the radius OA rotates through an angle θ, where

$$\omega = \frac{\theta}{t} \text{ and so } \theta = \omega t$$

But the arc length covered in this time, s, is given by

$$s = r\theta$$
and so $\quad\quad s = r\omega t$

The distance covered round the circular path is s in a time t. But the distance covered divided by the time taken is the linear speed v for the point on the rim of the wheel.   •

$$v = \frac{s}{t}$$
Hence $\quad\quad v = \frac{r\omega t}{t}$

$$v = r\omega \tag{8.5}$$

The linear velocity of a point on the rim of the wheel is numerically equal to the speed but is always directed along the tangent to the rim of the wheel.

Now consider the motion of the point on the rim of the wheel when the angular velocity is changing and there is an angular acceleration. The point on the rim of the wheel will then have a linear acceleration causing it to accelerate round its circular path. (Note that this is a linear tangential acceleration and is not the same as the radial acceleration which is referred to as the centripetal acceleration.) If, for fig. 8.3, the point on the rim has a linear velocity v_0 at point A and this has changed to v_1 by point B, then as this occurs in a time t the average linear acceleration a is given by

$$a = \frac{\text{change in linear velocity}}{\text{time taken}}$$

$$a = \frac{(v_1 - v_0)}{t}$$

If ω_0 is the angular velocity when the point is at A and ω_1 the angular velocity when it is at B, then using expression (8.5),

$$v_0 = r\omega_0 \text{ and } v_1 = r\omega_1$$

Hence $\qquad a = \dfrac{(r\omega_1 - r\omega_0)}{t}$

$$a = r\frac{(\omega_1 - \omega_0)}{t}$$

But the angular acceleration α is $(\omega_1 - \omega_0)/t$, hence
$$a = r\alpha \qquad\qquad\qquad (8.6)$$

Example 8.5 *A grinding wheel has a radius of 100 mm and rotates at 30 rev/s. What is the grinding speed at the circumference of the wheel?*

Since $\omega = 2\pi n$
$$\omega = 2\pi \times 30\,[\text{rev/s}]$$
$$\omega = 60\pi\,\text{rad/s}$$
As $\quad v = r\omega$
$$v = 0.100\,[\text{m}] \times 60\pi\,[\text{rad/s}]$$
$$v = 18.8\,\text{m/s}$$

Example 8.6 *The angular velocity of a car wheel increases from 5 rad/s to 50 rad/s in 30 s. If the wheel has a radius of 350 mm, what is (a) the* *average angular acceleration and* (b) *the average linear acceleration of a point on the rim of the wheel?*

(a) Since $\qquad \alpha = \dfrac{\omega_1 - \omega_0}{t}$

$$\alpha = \frac{(50 - 5)}{30\,[\text{s}]}[\text{rad/s}]$$

$$\alpha = 1.5\,\text{rad/s}^2$$

(b) Since $\qquad a = r\alpha$

$$a = 0.350\,[\text{m}] \times 1.5\,[\text{rad/s}^2]$$

$$a = 0.525\,\text{m/s}^2$$

8.5 Equations of angular motion

For motion with constant angular acceleration we can write a number of equations, comparable to the equations in chapter 6 for linear motion. Equation (8.4) can be rewritten in the form

$$\omega_1 = \omega_0 + \alpha t \qquad (8.7)$$

During the time interval t the average angular velocity is

$$\text{average angular velocity} = \frac{\omega_0 + \omega_1}{2}$$

Since, according to equation (8.2),

$$\text{average angular velocity} = \frac{\theta}{t}$$

then $\qquad \dfrac{\theta}{t} = \dfrac{\omega_0 + \omega_1}{2}$

$$\theta = \tfrac{1}{2}(\omega_0 + \omega_1)t$$

Since $\omega_1 = \omega_0 + \alpha t$ (equation 8.7), then

$$\theta = \tfrac{1}{2}(\omega_0 + \omega_0 + \alpha t)t$$
$$\theta = \omega_0 t + \tfrac{1}{2}\alpha t^2 \qquad (8.8)$$

If we take equation (8.7) and square it, we have

$$\omega_1^2 = (\omega_0 + \alpha t)^2$$

$$\omega_1^2 = \omega_0^2 + 2\omega_0\alpha t + \alpha^2 t^2$$

$$= \omega_0^2 + 2\alpha\,(\omega_0 t + \tfrac{1}{2}\alpha t^2)$$

122

Hence, using equation (8.8),

$$\omega_1^2 = \omega_0^2 + 2\alpha\theta$$

$$(8.9)$$

The following table show a comparison of the linear motion and the angular motion equations and quantities.

Linear motion	Angular motion
distance s	angle θ
velocity v	angular velocity ω
acceleration a	angular acceleration α
$v = s/t$ (6.1)	$\omega = \theta/t$ (8.2)
$v = u + at$ (6.2)	$\omega_1 = \omega_0 + \alpha t$ (8.7)
$s = ut + \frac{1}{2}at^2$ (6.3)	$\theta = \omega_0 t + \frac{1}{2}\alpha t^2$ (8.8)
$v^2 = u^2 + 2as$ (6.4)	$\omega_1^2 = \omega_0^2 + 2\alpha\theta$ (8.9)

Example 8.7 *A wheel, initially at rest, is subjected to a constant angular acceleration of 2.0 rad/s² for 50 s, calculate the angular velocity attained and the number of revolutions the wheel makes in that time.*

Since $\omega_1 = \omega_0 + \alpha t$
and $\omega_0 = 0$, then
$$\omega_1 = 0 + 2\cdot0\,[\text{rad/s}^2] \times 50\,[\text{s}]$$
$$= 100\,\text{rad/s}$$
Since $\theta = \omega_0 t + \frac{1}{2}\alpha t^2$
$$\theta = 0 + \frac{1}{2} \times 2\cdot0\,[\text{rad/s}^2] \times 50^2\,[\text{s}]^2$$
$$= 2500\,\text{rad}$$

One revolution is 2π rad, therefore the number of revolutions n is

$$n = \frac{2500\,[\text{rad}]}{2\pi\,[\text{rad}]}$$

$$= 397\cdot9$$

Example 8.8 *A wheel initially has an angular velocity of 50 rads/s, brakes are then applied and the wheel comes to rest in 15 s. What is the average retardation?*

Note that in the same way that retardation is applied to linear motion and means a negative acceleration, i.e. one in which the final velocity is less than the initial one, so the same significance can be applied to the term in angular motion, i.e. it is a negative angular acceleration.

Since $\omega_1 = \omega_0 + \alpha t$

$$0 = 50\,[\text{rad/s}] + \alpha \times 15\,[\text{s}]$$

$$\alpha = -\frac{50}{15} = -3{\cdot}3\ \text{rad/s}^2$$

8.6 Torque and angular motion

Torque, or the moment of a force, T about an axis is defined by

$$T = Fr \qquad (8.10)$$

where r is the radius of the turning circle to which the force is tangential. This is sometimes stated as torque is the product of the force and the perpendicular distance of the force from the axis of rotation.

Consider a rigid body which is rotating about an axis through O, as

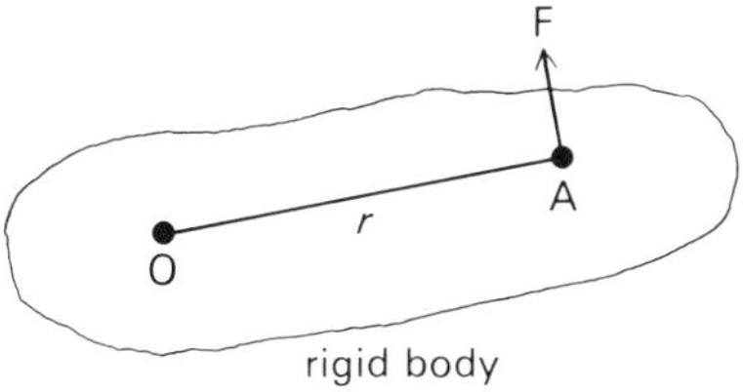

Fig. 8.4

in fig. 8.4. A particle of that body, at A, rotates through a turning circle of radius r. If F is the tangential force acting on this particle at A, then using Newton's Second law we have

$$F = ma$$

where a is the linear acceleration in the direction of the force. If α is the angular acceleration, then

$$a = r\alpha$$

and so $$F = mr\alpha$$

Hence the torque T acting on the particle is

$$T = Fr = mr^2\alpha$$

The rigid body is made up of a large number of small particles, each having different turning circles. The total torque acting on the body is the sum of the torques acting on all these small particles. Thus

Total torque = sum of all ($mr^2\alpha$) terms

All the particles will have the same angular acceleration. Thus

torque T = (sum of mr^2 terms) × α

The sum of all the mr^2 terms is called the moment of inertia, symbol I, of the body. Hence

$$T = I\alpha \qquad (8.11)$$

This equation for angular motion can be compared with the equation $F = ma$ for linear motion. Torque is the angular equivalent of force, angular acceleration is the angular equivalent of linear acceleration and moment of inertia is the angular equivalent of mass. Mass has been referred to, in section 1.7 of this book, as having a quality of inertia or reluctance to change velocity. Moment of inertia is a similar concept for angular motion, representing the inertia or reluctance to change angular velocity.

Example 8.9 *What torque has to be applied to a flywheel, with a moment of inertia of* 30 kg m², *to give it an angular acceleration of* 0·5 rad/s²*?*

$$\text{Since } T = I\alpha$$
$$T = 30\,[\text{kg m}^2] \times 0{\cdot}5\,[\text{rad/s}^2]$$
$$= 15\,\text{N m}$$

8.7 Moment of inertia

The *moment of inertia*, symbol I, of a body about a particular axis is defined by the equation

$$I = \Sigma mr^2 \qquad (8.12)$$

where r is the distance of a particle of mass m from the axis of rotation and the symbol Σ is used to indicate that the moment of inertia is the sum of all the mr^2 terms for all the particles in the body. Figure 8.5 shows the moments of inertia of some simple bodies.

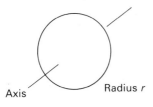
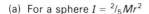

(a) For a sphere $I = {}^2/_5 Mr^2$

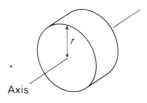

(b) For a disc $I = \frac{1}{2} Mr^2$

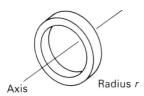

(c) For a ring $I = Mr^2$

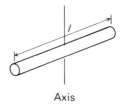

(d) For a slender rod $I = {}^1/_{12} Ml^2$

Fig. 8.5 Moments of inertia.

The moment of inertia of any body depends on the position of the axis about which the moment is considered. If the moment of inertia I_0 about a particular axis is known then the *theorem of parallel axes* enables the moment of inertia to be calculated for any parallel axis: if I is the moment of inertia about some parallel axis a distance d from the axis considered for the moment of inertia I_0 then

$$I = I_0 + Md^2 \qquad (8.13)$$

where M is the total mass of the body. Thus for a disc of radius r and mass M the moment of inertia about an axis through the centre is $\frac{1}{2} Mr^2$, the moment of inertia of the same disc about an axis tangential to the circumference of the disc is $\frac{1}{2} Mr^2 + Mr^2 = \frac{3}{2} Mr^2$.

Some objects though more complex than the objects shown in fig. 8.5 can be considered to be composed of a number of such simple objects. We could have, for instance, an object consisting of two spheres mounted at the ends of a slender rod, i.e. a dumb-bell. The moment of inertia of this object about an axis through the centre of the rod can be determined by adding together the moments of inertia of the component parts. The moment of inertia of a sphere about an axis through its centre is $\frac{2}{5}Mr^2$. What is required is the moment of inertia of the sphere about an axis through the centre of the rod. If the rod has a length l, i.e. the distance between the centres of the spheres is $l + 2r$ then the distance of the axis of rotation from the centre of a sphere is $\frac{1}{2}l + r$ and so the moment of inertia of a sphere about this axis is, using the parallel axis theorem, $\frac{2}{5} Mr^2 + M(\frac{1}{2}l + r)^2$. As there are two spheres the total moment of inertia due to the spheres is twice this value. The rod has also a moment of inertia of $\frac{1}{12}Ml^2$. Hence the total moment of inertia is

$$I = 2[\tfrac{2}{5}Mr^2 + M(\tfrac{1}{2}l + r)^2] + \tfrac{1}{12}Ml^2$$

Whatever the form of an object and however its mass is distributed, it is always possible to represent the moment of inertia in the form $I = mk^2$. Thus in the case of the disc where $I = \frac{1}{2}mr^2$ about the central axis, $k^2 = \frac{1}{2}r^2$. k is called the *radius of gyration*. The significance of this radius is that we can consider the body effectively to be behaving for rotation as though it had all its mass concentrated at a point a distance k from the axis of rotation.

Example 8.10 *A flywheel has a mass of 300 kg and a radius of gyration of* $1 \cdot 0$ m. *What is (a) the moment of inertia of the flywheel and (b) the torque necessary to give it an angular acceleration of* $0 \cdot 5$ rad/s^2?

(*a*)   Since $I = mk^2$

$$I = 300\,[\text{kg}] \times 1\cdot0^2\,[\text{m}]^2$$
$$= 300\,\text{kg m}^2$$

(*b*)   Since $T = I\alpha$

$$T = 300\,[\text{kg m}^2] \times 0\cdot5\,[\text{rad/s}^2]$$
$$= 150\,\text{N m}$$

Summary of Chapter 8

The unit of angular measurement is the radian, where one radian is the angle subtended at the centre of a circle by an arc equal in length to the radius.

$$\text{Arc length } s = r\theta \tag{8.1}$$
$$2\pi\,\text{rad} = 360°$$

Angular velocity is the rate of change of angular displacement with time.

$$\text{Average angular velocity } \omega = \frac{\theta}{t} \qquad (8.2)$$

$$\omega = 2\pi n \qquad (8.3)$$

where n is the number of revolutions per second.

Angular acceleration is the rate of change of angular velocity with time.

$$\text{Average angular acceleration } \alpha = \frac{(\omega_1 - \omega_0)}{t} \qquad (8.4)$$

The angular velocity ω is related to the linear velocity at a point distance r from the axis of rotation by

$$v = r\omega \qquad (8.5)$$

The angular acceleration is related to the linear acceleration by

$$a = r\alpha \qquad (8.6)$$

The equations of motion for motion with constant angular acceleration are

$$\omega_1 = \omega_0 + \alpha t \qquad (8.7)$$

$$\theta = \omega_0 t + \tfrac{1}{2}\alpha t^2 \qquad (8.8)$$

$$\omega_1^2 = \omega_0^2 + 2\alpha\theta \qquad (8.9)$$

Torque, or the moment of a force, T about an axis is defined by

$$T = Fr \qquad (8.10)$$

$$T = I\alpha \qquad (8.11)$$

The moment of inertia I about an axis is defined by

$$I = \Sigma mr^2 \qquad (8.12)$$

If I_0 is the moment of inertia about an axis then the moment of inertia I about a parallel axis a distance d away is given by

$$I = I_0 + Md^2 \qquad (8.13)$$

It is always possible to represent the moment of inertia of a body in the form $I = mk^2$, where k is the radius of gyration.

EXAMPLES 8

1. A flywheel rotates through 20 revolutions in 5 s. What is the average angular velocity?
2. A flywheel increases its rate of rotation from 1·0 to 2·0 rev/s in 12 s, what is the average angular acceleration?
3. A flywheel slows down from 10 rev/s to 4 rev/s in 100 s. What is the average angular retardation?
4. A grinding wheel rotates at 50 rev/s. Through what angle does a radius of the wheel rotate in 20 s?
5. A train has a constant linear velocity of 30 m/s. If the driving wheels have a radius of 1·0 m what is their angular velocity?
6. The wheels of a car increase their rate of rotation from 1·0 rev/s to 8·0 rev/s in 20 s. Calculate (a) the angular acceleration of a wheel and (b) the linear acceleration of a point on the rim of the wheel if the wheel has a radius of 350 mm.
7. A grinding wheel has a radius of 100 mm. What is the linear velocity of a point on the rim of the wheel when it is rotating at 50 rev/s?
8. A flywheel rotates through 80 revolutions in starting from rest and attaining a rate of revolution of 160 rev/min. What is (a) the angular velocity attained and (b) the average angular acceleration?
9. The armature of an electric motor is running at 100 rev/s. When the current is switched off the armature comes to rest in 12 s. What is (a) the average angular retardation and (b) the number of revolutions made by the armature in coming to rest?
10. A wheel initially has an angular velocity of 30 rad/s. When the brakes are applied the wheel comes to rest in 40 complete revolutions. What is (a) the average retardation and (b) the time taken for the wheel to come to rest?
11. What torque is required to produce an angular acceleration of 0·4 rad/s² in a wheel having a moment of inertia of 400 kg m²?
12. The rotor of an electric motor has a moment of inertia of 6·0 kg m². What torque is needed to accelerate the rotor at 0·2 rad/s²?
13. A flywheel has a mass of 300 kg and a radius of gyration of 0·5 m. Calculate (a) the moment of inertia and (b) the constant torque needed to bring such a wheel to rest in 30 s when it is rotating at 5·0 rev/s.
14. A flywheel has a mass of 60 kg and a radius of gyration of 360 mm. Calculate the torque needed to increase the rate of revolution from 4·0 to 5·0 rev/s in 30 s.

ANSWERS TO EXAMPLES 8

1. 25·1 rad/s
2. 0·52 rad/s²
3. 0·38 rad/s²
4. 6283·2 rad
5. 30 rad/s
6. (a) 2·20 rad/s², (b) 0·77 m/s².
7. 31·4 m/s
8. (a) 16·8 rad/s, (b) 1·76 rad/s².
9. (a) 52·4 rad/s², (b) 599·5.
10. (a) 70·7 rad/s², (b) 2·69 s.
11. 160 N m
12. 1·2 N m
13. (a) 75 kg m², (b) 78·5 N m
14. 1·63 N m

CHAPTER 9

Work, power and energy

9.1 Work

It was stated in section 1.9 that when a force is exerted against some
form of resistance through a distance in the direction of the force,
work is done, and that the SI unit of work is the *joule*, namely *the
work done when a force of* 1 *newton is exerted through a distance of*
1 *metre in the direction of the force.* Hence, if a force F, in newtons,
is exerted through a distance s, in metres, in the direction of the
force,

$$\text{work done} = F \text{ [newtons]} \times s \text{ [metres]}$$
$$= Fs \text{ joules} \tag{9.1}$$

Example 9.1 *A mass of* 40 *kg rests on a horizontal surface. If the coefficient of friction is* 0·2, *calculate the work done in moving this mass
through a distance of* 10 *m.*

Vertical force on surface $\simeq 40 \times 9\cdot81 = 392\cdot4$ N.

From section 5.1,

$$\left.\begin{array}{r}\text{horizontal force required} \\ \text{to overcome friction}\end{array}\right\} = 392\cdot4 \text{ [N]} \times 0\cdot2 = 78\cdot48 \text{ N,}$$

∴ $\qquad$ work done $= 78\cdot48$ [N] $\times 10$ [m] $= 784\cdot8$ J.

9.2 Work represented by an area or diagram of work

The amount of work done by a force exerted through a distance in
its own direction may be represented by a rectangular area or
diagram of work, one side of which represents the force to scale
and the other the distance to scale. For example, in fig. 9.1, if a
force of 50 N is represented on a scale of 1 mm to 1 N and moves
through a distance of 12 m represented on a scale of 1 mm to 0·2 m,
then 1 mm² of area represents

$$1 \text{ [N]} \times 0\cdot2 \text{ [m]} = 0\cdot2 \text{ J.}$$

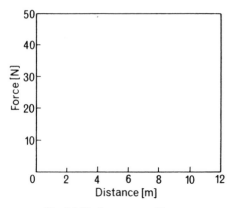

Fig. 9.1 Work represented by an area.

The sides of the rectangle are 50 mm by 60 mm so that the total area is 3000 mm^2.

Hence the work done is represented by an area of 3000 mm^2 on a scale of 1 mm^2 to 0·2 J,

∴   work done = 0·2 [J/mm^2] × 3000 [mm^2] = 600 J.

For calculating the work done by a force of constant value, the diagram is of no advantage; for instance, in the case just considered, we could have calculated the work done by merely multiplying the force by the distance thus:

work done = 50 [N] × 12 [m] = 600 J.

When the force varies in some known but perhaps rather complicated manner, calculation of the area of the work diagram may be a convenient way of determining the work done.

Example 9.2 *A spring, initially in a state of ease (neither stretched nor compressed), is extended 50 mm. Calculate the work done if the spring requires a force of 0·8 N per millimetre of stretch.*

Force for 50-mm extension = 50 [mm] × 0·8 [N/mm] = 40 N.

Fig. 9.2 shows how the extension increases in proportion to the force as the latter is increased from zero to 40 N, and the work done is represented by the area of the shaded triangle.

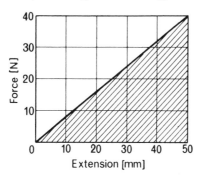

Fig. 9.2 Force/extension diagram for Example 8.2.

Alternatively, the work done can be calculated thus:

average force during extension $= 40/2 = 20$ N
and   total extension $= 50$ mm $= 0.05$ m,
$\therefore$ work done $=$ average force $\times$ extension
                             $= 20$ [N] $\times$ 0.05 [m] $= 1$ J.

Example 9.3 *A chain, 20 m long, has a mass of 12 kg/m length and hangs vertically. Draw the work diagram and determine (a) the work done in winding the chain on to a drum at the top and (b) the work done in winding the chain up the first 10 m.*

(a) Weight of chain per metre length $\simeq 12 \times 9.81 = 117.7$ N
$\therefore$ total weight of chain $= 117.7 \times 20 = 2354$ N.

Hence the total lifting force required at first is 2354 N and the lifting force diminishes uniformly to zero as the chain is being wound on the drum.

Fig. 9.3 is the diagram of work and shows how the lifting force varies with the length through which this force is being exerted at all stages of the winding process.

If the scales are 1 mm to 50 N for the lifting force and 1 mm to 0.5 m for the distance through which the chain is raised, then 1 mm² represents 50 N $\times$ 0.5 m, namely 25 J. The total area of the diagram, i.e. triangle AOB, is OB $\times$ mean height EC. But OB $= 40$ mm and EC $= 23.5$ mm,

$\therefore$ area of triangle AOB $= 40 \times 23.5 = 940$ mm² and

$$\text{work done} = 25 \, [\text{J/mm}^2] \times 940 \, [\text{mm}^2]$$
$$= 23\,500 \text{ J} = 23 \cdot 5 \text{ kJ}.$$

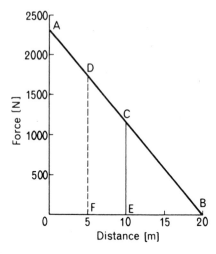

Fig. 9.3 Work done by a uniformly varying force.

In such a simple case, it is not necessary to calculate the area and the number of joules per square millimetre. The diagram shows that the *average* force, represented by EC, is half the maximum force OA which is 2354 N; hence,

$$\text{work done} = \text{average lifting force} \times \text{distance}$$
$$= \tfrac{1}{2} \times 2354 \, [\text{N}] \times 20 \, [\text{m}]$$
$$= 23\,540 \text{ J} = 23 \cdot 54 \text{ kJ}.$$

The discrepancy between the two values is due to the approximation in reading the value of EC from the graph.

(b) Similarly, though the work done in the first 10 m of lift is represented by the area ACEO, it is not necessary to calculate the area. It is evident from fig. 9.3 that:

$$\text{FD} = \tfrac{3}{4} \times \text{OA} = \tfrac{3}{4} \times 2354 = 1765 \text{ N}$$
$$= \text{average force during first 10 m of lift,}$$

$\therefore$　$\left.\begin{array}{r}\textbf{work done during first} \\ \textbf{10 m of lift}\end{array}\right\} = 1765 \, [\text{N}] \times 10 \, [\text{m}]$

$$= 17\,650 \text{ J} = 17 \cdot 65 \text{ kJ}.$$

Example 9.4 *A load is hauled with a tractive effort F, in newtons, which varies with the distance x, in metres, moved as shown in the following table:*

x [m]	0	20	50	80	110	130	160	190	200
F [N]	1280	1270	1220	1110	905	800	720	670	660

Determine the total work done in kilojoules.

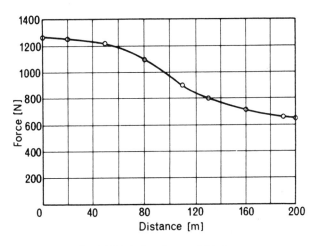

Fig. 9.4 Work done by a variable force.

First plot the graph showing the values of *F* for different values of *x* (as in fig. 9.4), using scales of, say, 1 mm to 2 m for *x* and 1 mm to 20 N for *F*.

The average tractive effort can be determined by means of the mid-ordinate rule, i.e. the height is measured at the middle of each of 10 vertical strips into which the diagram is divided. The sum of these average heights divided by 10 gives the average height of the whole graph. The following table gives the tractive effort, derived from the graph of fig. 9.4, at the mid-point of each 20-m strip:

x [m]	10	30	50	70	90	110	130	150	170	190
F [N]	1275	1260	1220	1150	1050	905	800	740	695	670

The sum of the 10 values of F is 9765 N,

hence   average tractive effort = $9765/10 = 976.5$ N

and   work done = 976.5 [N] $\times$ 200 [m]

                                                     = 195 300 J = 195·3 kJ.

9.3 Work done by an oblique force

In section 2.9 it was shown that when a force F, acting on a body, is inclined at an angle θ to the direction of motion (fig. 9.5), the force can be resolved into two components, namely $F\cos\theta$ acting in the direction of motion and $F\sin\theta$ acting at right angles to the direction

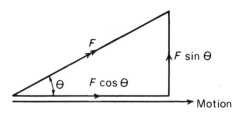

Fig. 9.5 Work done by an oblique force.

of motion. Since there is no movement of the body in the direction of component $F\sin\theta$, the latter does not do any work. Hence the work done by the oblique force F is the product of the first component, $F\cos\theta$, and the distance s through which the body moves:

i.e.   work done $= (F\cos\theta) \times s$.

Example 9.5 *A barge is being towed along a canal at* 4 *km/h. The tow rope is inclined at an angle of* 25° *to the direction of motion of the barge, and the pull on the rope is* 300 *N. Calculate the work done in* 2 *minutes.*

$$4 \text{ km/h} = 4000 \text{ [m]}/3600 \text{ [s]} = 1.111 \text{ m/s}$$

$\therefore$ distance travelled $\Big\}$
          in 2 minutes $= 1.111$ [m/s] $\times (2 \times 60)$ [s] $= 133.3$ m.

$$\left.\begin{array}{r}\text{Component of pull in}\\ \text{direction of force}\end{array}\right\} = 300 \cos 25° = 300 \times 0.9063$$

$$= 271.9 \text{ N},$$

$$\therefore \quad \text{work done} = 271.9 \text{ [N]} \times 133.3 \text{ [m]}$$

$$= 36\,250 \text{ J} = 36.25 \text{ kJ.}$$

9.4 Work done in rotation

Let us consider the case of a crank handle or a pulley attached to a shaft, as in fig. 9.6, and suppose a force of 70 N to be exerted at right angles to the crank arm, 200 mm long, or at the circumference of the pulley of 200 mm radius, in order to turn the shaft against some resistance. Also, suppose the force to be always in the direction of the motion of the point to which it is applied.

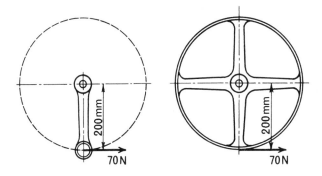

Fig. 9.6 Work done in rotation.

The distance through which the point of application of the force travels in 1 revolution is $2\pi \times 0.2$ [m], namely 1.257 m,

$$\therefore \quad \text{work done in 1 revolution} = 70 \text{ [N]} \times 1.257 \text{ [m]} = 88 \text{ J.}$$

In general, if a force F newtons acts at a radius r metres,

$$\text{work done in 1 revolution} = F \times 2\pi r \text{ joules} \qquad (9.2)$$

The product of a turning force and the radius of the circle at which it acts is termed the *torque* or *turning moment* about the axis of rotation. Hence, for a force F newtons acting at a radius r metres,

$$\text{torque (or turning moment)} = T = Fr \text{ newton metres}$$

and from expression (9.2),

$$\left.\begin{array}{l}\text{work done, in joules,} \\ \text{in } n \text{ revolutions}\end{array}\right\} = Fr \times 2\pi n$$

$$= \text{torque in newton metres} \\ \times \text{ angle in radians}$$

$$= T\theta \hspace{3cm} (9.3)$$

Example 9.6 *An electric motor has to lift a body having a mass of 50 kg by means of a rope wound round a drum having a diameter of 1·2 m. Calculate* (a) *the torque to be exerted and* (b) *the work done when the drum makes 20 revolutions.*

(a)   Force on rope $\simeq$ 50 × 9·81 = 490·5 N

and radius of drum = 0·6 m,

∴   torque = 490·5 [N] × 0·6 [m]

$$= 294\cdot3 \text{ N m.}$$

(b) From expression (9.3),

work done = 294·3 [N m] × (2π × 20) [rad]

$$= 37\,000 \text{ J} = 37 \text{ kJ.}$$

Example 9.7 *A pulley is 800 mm in diameter and the difference in tension on the two sides of the driving belt is 2000 N. If the speed of the pulley is 300 rev/min, what is the work done in 5 min?*

Radius of pulley = 400 mm = 0·4 m,

∴   torque = 2000 [N] × 0·4 [m]

$$= 800 \text{ N m.}$$

No. of revolutions in 5 minutes = 300 × 5 = 1500.

From expression (9.3),

work done = 800 [N m] × (2π × 1500) [rad]

$$= 7\,540\,000 \text{ J} = 7\cdot54 \text{ MJ.}$$

9.5 Power

Power is the rate of doing work and the SI unit of power is the *watt*, namely 1 joule per second. In practice, the watt is often found to be inconveniently small; consequently the *kilowatt* (kW) is frequently used, the kilowatt being 1000 watts. For still larger powers, the *megawatt* (MW) is used, where:

$$1 \text{ MW} = 1000 \text{ kW} = 1\,000\,000 \text{ W}.$$

Similarly, when we are dealing with a large amount of work (or energy), it is often convenient to express the latter in *kilowatt hours*.

$$\begin{aligned}1 \text{ kW h} &= 1000 \text{ watt hours} \\ &= 1000 \times 3600 \text{ watt seconds or joules} \\ &= 3\,600\,000 \text{ J} = 3\cdot6 \text{ MJ}.\end{aligned}$$

9.6 Power required for rotation

If T be the torque or turning moment, in newton metres, and if n be the speed, in revolutions per second, then from expression (9.3) we have:

$$\begin{aligned}\text{work done per second} &= \text{torque in newton metres} \\ &\quad \times \text{ speed in radians per second} \\ &= T \text{ [newton metres]} \\ &\quad \times 2\pi n \text{ [radians/second]} \\ &= 2\pi n T \text{ joules/second or watts}\end{aligned}$$

i.e.
$$\begin{aligned}\text{power} &= 2\pi n T \text{ watts} \\ &= \omega T \text{ watts}\end{aligned} \qquad (9.4)$$

where
$$\begin{aligned}\omega &= \text{angular velocity in radians/second} \\ &= 2\pi n \text{ radians/second}.\end{aligned}$$

If the rotational speed be N revolutions per minute,

$$\text{power} = 2\pi T N/60 \text{ watts} \qquad (9.5)$$

Example 9.8 *A motor vehicle hauls a trailer at 75 km/h when exerting a steady pull of 800 N. Calculate (a) the work done in 20 min (i) in megajoules and (ii) in kilowatt hours, (b) the power required.*

(*a*) (*i*)
$$\begin{aligned}\text{Distance travelled} &= 75 \text{ [km/h]} \times (20/60) \text{ [h]} \\ &= 25 \text{ km,}\end{aligned}$$

∴
$$\begin{aligned}\text{work done} &= 800 \text{ [N]} \times (25 \times 1000) \text{ [m]} \\ &= 20\,000\,000 \text{ J} = 20 \text{ MJ.}\end{aligned}$$

(*ii*)
$$\begin{aligned}&\text{Since } 1 \text{ kW h} = 3\cdot6 \text{ MJ} \\ &\text{work done} = 20/3\cdot6 = 5\cdot56 \text{ kW h.}\end{aligned}$$

(b)
$$\text{Power} = \frac{\text{work done in joules}}{\text{time in seconds}}$$

$$= \frac{20\ 000\ 000\ [\text{J}]}{(20 \times 60)\ [\text{s}]} = 16\ 670\ \text{W}$$

$$= 16 \cdot 67\ \text{kW}.$$

Alternatively,

$$\text{power} = \frac{\text{work done in kilowatt hours}}{\text{time in hours}}$$

$$= \frac{5 \cdot 56\ [\text{kW h}]}{(20/60)\ [\text{h}]} = 16 \cdot 68\ \text{kW}.$$

Example 9.9 *An electric motor is developing 8 kW at a speed of 1200 rev/min. Calculate (a) the work done in 45 min (i) in kilowatt hours, (ii) in megajoules, and (b) the torque in newton metres.*

(a) (i)
$$\text{Work done} = 8\ [\text{kW}] \times (45/60)\ [\text{h}]$$
$$= 6\ \text{kW h}.$$

(ii)   Since 1 kW h = 3·6 MJ,

∴   work done = 6 × 3·6 = 21·6 MJ.

(b) From expression (9.5),

$$8000\ [\text{W}] = 2\pi T \times (1200/60)\ [\text{rev/s}]$$

∴ $T = 63 \cdot 7$ N m.

Example 9.10 *The tensions on the two sides of a belt passing round a pulley are 2200 N and 460 N respectively. The effective diameter of the pulley is 400 mm and the speed is 700 rev/min. Calculate (a) the power transmitted and (b) the work done in 10 min (i) in megajoules and (ii) in kilowatt hours.*

(a) The effective driving force of a belt is the difference between the tensions on the tight and the slack sides,

∴   net driving force = 2200 − 460 = 1740 N.

Effective radius of pulley = 200 mm = 0·2 m,

∴   torque = 1740 [N] × 0·2 [m] = 348 N m.

From expression (9.5), we have:

$$\text{power} = 2\pi \times 348 \,[\text{N m}] \times (700/60) \,[\text{rev/s}]$$
$$= 25\,500 \text{ W} = 25 \cdot 5 \text{ kW.}$$

(b) (i)   Work done in 10 min $= 25\,500 \,[\text{W}] \times (10 \times 60) \,[\text{s}]$
$$= 15\,300\,000 \text{ J} = 15 \cdot 3 \text{ MJ.}$$

(ii)   Work done in 10 min $= 25 \cdot 5 \,[\text{kW}] \times (10/60) \,[\text{h}]$
$$= 4 \cdot 25 \text{ kW h.}$$

Alternatively, since 1 kW h $= 3 \cdot 6$ MJ

∴   work done in 10 min $= 15 \cdot 3/3 \cdot 6 = 4 \cdot 25$ kW h.

9.7 Determination of the output power of a machine by means of a brake

In the case of a comparatively small machine, the output power can be measured by some form of mechanical brake such as that shown in fig. 9.7, where a belt (or rope) on an air or water-cooled pulley has its ends attached to spring balances S_1 and S_2, calibrated in newtons. The balances are supported by a rigid horizontal beam B, and the tension on the belt can be controlled by wing-nuts W.

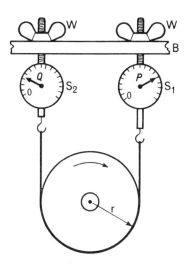

Fig. 9.7 Brake test.

Suppose the brake pulley to be rotating clockwise and the tension on the belt to be adjusted to give readings of P and Q newtons on S_1 and S_2 respectively. The pull P exerted by S_1 has to balance the pull Q exerted by S_2 and the friction force F between the belt and the pulley,

i.e. $$P = Q + F$$
∴ $$F = (P - Q) \text{ newtons.}$$

If r be the effective radius of the brake, in metres,

torque due to brake friction $= Fr$
$$= (P - Q)r \text{ newton metres.}$$

If N be the speed of the pulley in revolutions/minute, then from expression (9.5),

$$\text{output power} = 2\pi(P - Q)rN/60 \tag{9.6}$$

The output power of the machine is converted into heat at the brake, and the size of the machine that can be tested by this method is limited by the difficulty of dissipating this heat.

9.8 Efficiency of a Machine

In all machines, some of the power supplied to the machine is lost in overcoming friction, etc., so that the useful power available is less than the input power. The ratio of the output power to the input power is termed the *efficiency* of the machine,

i.e. $$\text{efficiency} = \frac{\text{output power}}{\text{input power}}$$

and is expressed as a 'per unit' value or as a percentage; thus, if the input and output powers are 75 kW and 60 kW respectively,

$$\text{efficiency} = 60/75 = 0\cdot8 \text{ per unit}$$
$$= 0\cdot8 \times 100 = 80 \text{ per cent.}$$

The input and output powers must obviously be expressed in the same unit.

Example 9.11 *In a brake test on an electric motor, the readings on balances S_1 and S_2 (fig. 9.7) were 440 N and 87 N respectively. The* *effective diameter of the pulley was 500 mm and the speed was 800 rev/min. The power supplied to the motor was 8·5 kW. Calculate* (a) *the output power of the motor and* (b) *the efficiency.*

(a)   Effective radius of pulley = 250 mm = 0·25 m.

Net pull due to friction = 440 − 87 = 353 N,

∴   torque = 353 [N] × 0·25 [m] = 88·25 N m.

Substituting in expression (9.5), we have:

output power = 2π × 88·25 [N m] × (800/60) [rev/s]
= 7390 W = 7·39 kW.

(b)   Efficiency = 7·39 [kW]/8·5 [kW] = 0·87 per unit
= 87 per cent.

9.9 Energy

When a body is capable of doing work, it is said to possess *energy* which may take various forms such as mechanical energy, thermal energy, chemical energy and electrical energy. In mechanics we are concerned only with mechanical energy which is of two kinds, namely *kinetic energy* and *potential energy*.

The kinetic energy of a body is the energy it possesses by virtue of its motion. Thus a body, set in motion by a force doing work upon it, acquires kinetic energy which enables it to do work against resisting forces. An important engineering application is the flywheel. Work is done on the flywheel as its speed is increased; and later, when the machine to which it is attached slows down, some of the stored kinetic energy is given out by the flywheel and helps to drive the machine. This is the reason why a machine driving a fluctuating load, such as a stamping press, is usually fitted with a flywheel to maintain a more constant speed than would otherwise be the case.

If a force is exerted on a body and there is no opposing resistance except the inertia of the body, the whole of the work done becomes the kinetic energy of the body. Thus, if a force F, acting on a mass m, gives it a uniform acceleration a, then from expression (7.2):

$$F = ma$$

If s is the distance travelled by the body while it accelerates from standstill to a velocity v, then from expression (6.4), we have:

$$s = \tfrac{1}{2} v^2 / a$$

$$\therefore \quad \text{work done} = F \text{ [newtons]} \times s \text{ [metres]}$$
$$= ma \times \tfrac{1}{2} v^2 / a$$
$$= \tfrac{1}{2} mv^2 \text{ joules} \qquad (9.7)$$
$$= \text{kinetic energy of body.}$$

The potential energy of a body is the energy it possesses by virtue of its position or state of strain. For instance, a body raised to a height above the ground has potential energy since its weight can do work as the body returns to the ground.

If a body having a mass m, in kilograms, is lifted vertically through a height h, in metres, and if g is the gravitational acceleration in metres/second² at that point,

$$\text{force required} = \text{weight of the body}$$
$$= mg \text{ newtons}$$
$$\text{and} \qquad \text{work done} = \text{weight of the body} \times \text{height}$$
$$= mgh \text{ joules} \qquad (9.8)$$
$$= \text{potential energy of body.}$$

Another illustration of the storage of potential energy is given in Example 9.2, where a spring, initially at ease, i.e. in its zero position, is extended 50 mm, thereby storing 1 joule of potential energy (sometimes referred to as *elastic strain energy*) in the spring.

The pendulum of a clock is an example of energy being changed backwards and forwards between the kinetic form and the potential form. Thus the oscillating mass has its maximum kinetic energy at the lowest point of its travel, its potential energy being then zero. On the other hand, the potential energy of the pendulum is a maximum at the end of each swing, its speed and therefore its kinetic energy being then zero. The small loss of energy due to friction is supplied by the impulse given regularly through the escapement mechanism from the mainspring.

One of the natural sources of potential energy is water lifted by evaporation from sea-level to lakes and rivers at higher levels into which it is deposited as rain or snow. With the aid of pipes supplying water turbines at a lower level, the potential energy of the water stored in a reservoir at a high level can be converted into kinetic energy and thereby used to drive the turbines which in turn drive electrical generators or other machinery. Thus the potential energy of the water in the reservoir is converted into useful work.

9.10 Principle of the conservation of energy

This important principle states that whenever energy is converted from one form to another, no energy is lost. All the energy involved in the conversion can be accounted for in some form or another. This constancy of the total energy is referred to as the *Principle of the Conservation of Energy*.

In some situations we can have, to a reasonable approximation, a conservation of mechanical energy. Thus with the swinging pendulum of the clock potential energy at the end of each swing is converted to kinetic energy at the lowest point of its travel. The sum of the potential and kinetic energies is a constant.

Example 9.12 *A body having a mass of 30 kg is supported 50 m above the earth's surface. What is its potential energy?*

If the body is allowed to fall freely, calculate its potential and kinetic energies (a) *when the body is 20 m above the ground and* (b) *just before it touches the ground.*

$$\text{Weight of body} \simeq 30 \times 9\!\cdot\!81 = 294\!\cdot\!3 \text{ N,}$$

$$\therefore \quad \left.\begin{array}{c}\text{work done in lifting} \\ \text{the body 50 m}\end{array}\right\} = 294\!\cdot\!3 \text{ [N]} \times 50 \text{ [m]}$$

$$= 14\ 715 \text{ J} = 14\!\cdot\!715 \text{ kJ,}$$

i.e. potential energy of the body when it is 50 m above ground is 14·715 kJ. Since the body is stationary, its kinetic energy is zero.

(*a*) When the body is 20 m above ground,

$$\text{its potential energy} = 294\!\cdot\!3 \text{ [N]} \times 20 \text{ [m]}$$

$$= 5886 \text{ J} = 5\!\cdot\!886 \text{ kJ.}$$

Vertical distance travelled by body = 50 − 20 = 30 m.

Since $$v = \sqrt{(2gs)} \qquad (6.7)$$

and assuming $$g = 9\!\cdot\!81 \text{ m/s}^2, \text{ we have:}$$

$$v = \sqrt{(2 \times 9\!\cdot\!81 \times 30)} = 24\!\cdot\!26 \text{ m/s.}$$

From expression (9.7),

$$\text{kinetic energy} = \tfrac{1}{2} \times 30 \text{ [kg]} \times (24\cdot26)^2 \text{ [m/s]}^2$$
$$= 8829 \text{ J} = 8\cdot829 \text{ kJ}.$$

Alternatively, from the principle of the conservation of energy, the sum of the potential and kinetic energies must remain constant since no energy is being converted into any other form of energy; hence:

$$\text{kinetic energy} = \text{potential energy at 50 m}$$
$$- \text{ potential energy at 20 m}$$
$$= 14\cdot715 - 5\cdot886 = 8\cdot829 \text{ kJ}.$$

(*b*)   Velocity of body just $\left.\begin{array}{l}\\ \\\end{array}\right\}$ before it touches ground $= \sqrt{(2 \times 9\cdot81 \times 50)} = 31\cdot32 \text{ m/s}$

$\therefore$ $$\text{kinetic energy} = \tfrac{1}{2} \times 30 \text{ [kg]} \times (31\cdot32)^2 \text{ [m/s]}^2$$
$$= 14\,715 \text{ J} = 14\cdot715 \text{ kJ},$$

i.e. the whole of the initial potential energy has been converted into kinetic energy. When the body is finally brought to rest by the resistive force of the ground, practically the whole of this energy is converted into heat.

Example 9.13 *A motor vehicle having a mass of 2 Mg is travelling along a straight level road at 60 km/h. What is its kinetic energy in kilojoules?*

If the resistance to motion remains constant at 14 newtons per kilonewton weight of the vehicle, how far will it travel, without driving force or brakes, before coming to rest?

$$60 \text{ km/h} = 60\,000 \text{ [m]}/3600 \text{ [s]} = 16\cdot67 \text{ m/s}.$$

Mass of vehicle $= 2 \text{ Mg} = 2000 \text{ kg}$,

$\therefore$ $$\text{kinetic energy} = \tfrac{1}{2} \times 2000 \text{ [kg]} \times (16\cdot67)^2 \text{ [m/s]}^2$$
$$= 278\,000 \text{ J} = 278 \text{ kJ}.$$

Weight of vehicle $\simeq 2000 \times 9\cdot81 = 19\,620 \text{ N}$
$$= 19\cdot62 \text{ kN},$$

$\therefore$ resistance to motion $= 19\cdot62 \text{ [kN]} \times 14 \text{ [N/kN]}$
$$= 275 \text{ N}.$$

Since the whole of the kinetic energy of 278 000 J has to be expended in bringing the vehicle to rest,

$\therefore$ $$278\,000 \text{ [J]} = 275 \text{ [N]} \times \text{distance in metres}$$
so that $$\text{distance} = 1010 \text{ m} = 1\cdot01 \text{ km}.$$

9.11 Impact tests

Impact tests are designed to simulate the response of a material to high rates of loading. They involve the test piece being struck a sudden blow. There are two main forms of the test used with metals, the Izod and the Charpy tests. Both tests involve the imparting of a blow to the testpiece through a striker mounted on a pendulum, fig. 9.8 showing the basic principle. The energy absorbed in breaking the test piece is measured. This is done by allowing the pendulum striker to fall from a standard height and then measuring the height to which the striker rises on the other side of its swing after breaking through the test piece.

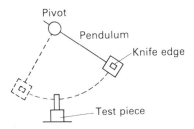

Fig. 9.8 The principle of impact testing.

If the initial height of the striker above the point of impact with the test piece is h_0 then it has a potential energy of mgh_0. When it falls, this potential energy is converted into kinetic energy. At the point of impact all this potential energy will be kinetic energy. After the impact the kinetic energy which the striker has left is converted into potential energy and the striker climbs up to a height h. The potential energy is then mgh. Thus the energy absorbed by the test piece being broken is $(mgh_0 - mgh)$.

The difference between the two forms of test is the form of the test piece and how the blow is struck. Figure 9.9 shows the forms of the test pieces and where the striker hits them. Both test pieces are notched. The Izod test piece is mounted as a cantilever and the blow struck on the same face as the notch and at a fixed height above it. The Charpy test piece is mounted as a beam and the blow struck on the opposite face to the notch, directly opposite the notch. The form of the notch is important and standard notches of different forms are used. The form of the notch used needs to be quoted, along with whether it is Izod or Charpy, in the result.

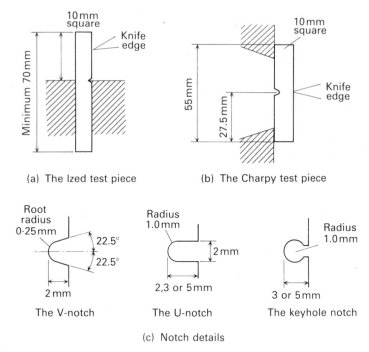

(a) The Ized test piece   (b) The Charpy test piece

(c) Notch details

Fig. 9.9 The impact test pieces.

The way in which a metal fractures can be considered to fall into two classes, a brittle fracture or a ductile fracture. With a brittle fracture there is little plastic deformation during the propagation of the crack through the material. With a ductile fracture there is considerable plastic deformation. The greater the amount of plastic deformation the greater the energy needed for the failure. Thus the impact test data gives an indication of the brittleness of materials.

A particular use of impact tests is to determine whether heat treatment has been correctly carried out in that a comparatively small change in heat treatment can lead to quite noticeable changes in impact test results.

9.12 Other forms of energy

There are other forms of energy such as *thermal* energy, *electrical* energy and *chemical* energy. Thermal energy can be produced by doing mechanical work; for example, when a piece of metal is hammered vigorously, the gain of thermal energy may be sufficient

to be detected by the sense of touch. Thermal energy is also produced from mechanical work when the surface of one body rubs or
slides over the surface of another body, particularly when the
surfaces are rough.

Electrical energy can be generated from the mechanical energy
developed by an engine or a turbine driving an electric generator.
Conversely, electrical energy supplied to an electric motor enables
the latter to supply mechanical energy to a machine coupled to its
shaft.

Chemical energy can be produced from electrical energy as, for
example, when a secondary cell (section 21.4) is charged. Conversely, when the cell is discharged, chemical energy is converted
into electrical energy. The latter can be converted into thermal and
light energies when a suitable lamp is connected across the cell, or
into mechanical energy if the cell is used to drive an electric motor.

In some situations we can have, to a reasonable approximation, a
conservation of mechanical energy. Thus with the swinging pendulum of the clock potential energy at the end of each swing is converted to kinetic energy at the lowest point of its travel. The sum of the
potential and kinetic energies is a constant.

Summary of Chapter 9

Work is done when force is exerted through a distance in its own
direction. Energy is thus converted to or from some form of
mechanical energy including kinetic energy and potential energy;
the latter may be either gravitational or elastic strain energy. Or it
may be converted into thermal energy, electrical energy or chemical
energy. But in any transformation, no energy is destroyed (the
Principle of Conservation of Energy).

Work is the product of force and the distance moved in the
direction of the force, or the product of the torque and the angle of
rotation in radians. The SI unit of work is the *joule*.

Power is the rate of doing work and the SI unit of power is the
watt, namely 1 joule/second.

The symbol for work and energy is W and that for power is P.

$$W \text{ [joules]} = F \text{ [newtons]} \times s \text{ [metres]} \qquad (9.1)$$

$$= T \text{ [newton metres]} \times \theta \text{ [radians]} \qquad (9.3)$$

$$1 \text{ kW h} = 3\ 600\ 000 \text{ J} = 3 \cdot 6 \text{ MJ}.$$

Work done by an oblique force $= Fs \cos \theta$

$$\text{Power [watts]} = \omega \text{ [rad/s]} \times T \text{ [N m]} \tag{9.4}$$
$$= 2\pi T \text{ [N m]} \times (N/60) \text{ [rev/s]} \tag{9.5}$$

The energy of a body is its capacity for doing work.

Kinetic energy is the energy possessed by a body by virtue of its motion.

$$\text{Kinetic energy, in joules} = \tfrac{1}{2} m \text{ [kg]} \times v^2 \text{ [m/s]}^2 \tag{9.7}$$

Potential energy is the energy possessed by a body by virtue of its position or state of strain.

For a body of mass m lifted through height h,

$$\text{potential energy, in joules} = m \text{ [kg]} \times g \text{ [m/s}^2] \times h \text{ [m]} \tag{9.8}$$
$$\simeq 9 \cdot 81 \ mh.$$

$$\text{Efficiency} = \frac{\text{output power}}{\text{input power}}$$

EXAMPLES 9

1. The work done in moving a body through a distance of 100 m in the direction of the force is 750 J. Calculate the value of the force.
2. Calculate the work done, in kilojoules, in lifting a mass of 800 kg through a vertical height of 40 m.
3. A force of 250 N is applied to a body. If the work done is 30 kJ, what is the distance through which the body is moved?
4. The work done in lifting a body vertically through a distance of 120 m is 5 W h. Calculate the mass of the body.

   If the time taken is 2 min, what is the average value of the power?
5. A force F newtons acting on a body in the direction of its motion varies as follows for different distances s metres from the initial position:

F	200	360	450	480	400	300
s	0	10	20	30	40	50

   Draw the work diagram to scale and determine (a) the average force and (b) the work done, in kilojoules, when the body is moved through a distance of 50 m. Assume the points on the graph to be joined by straight lines.
6. A car has the following tractive forces, in kilonewtons, exerted on it after it has travelled distances s metres from rest:

F	6·4	6·2	5·6	4·8	4·1	3·6	3·3
s	0	5	10	15	20	25	30

   Determine the work done (a) in kilojoules and (b) in watt hours when the car has moved through 30 m.

If the time taken to travel 30 m is 20 s, what is the average value of the power in kilowatts?

7. A helical spring is extended by a force which increases uniformly from zero to 600 N. The corresponding extension of the spring is 200 mm. Draw a work diagram and determine the total work done.

8. A compression spring is 150 mm long when unloaded and 120 mm long when subjected to a force of 20 N. It is 80 mm long when fully compressed. Draw a diagram showing the relationship between load and length of spring between zero and full load. Determine (a) the force to produce full compression and (b) the work done in compressing the spring between 150 mm and 80 mm and (c) the work done in compressing the spring between 120 mm and 80 mm.

9. A chain, having a mass of 18 kg/m of length, is 30 m long and hangs vertically. How much work is done in upwinding the chain on to a drum?

10. A chain, 200 m long, having a mass of 12 kg/m and hanging vertically, is wound up on to a drum. Draw to scale a graph showing the lifting forces as ordinates and the lengths drawn up, from 0 to 200 m, as abscissae. From the graph calculate the work done in winding up (a) the first 80 m of the chain and (b) the whole chain.

11. Calculate the work done (a) in megajoules and (b) in kilowatt hours, in emptying a circular shaft, 4 m diameter and 60 m deep, which is full of water, assuming the water flows away at the top of the shaft. Neglecting losses, calculate the time taken by a pump, developing 4 kW, to empty the shaft. Assume 1 m³ of water to have a mass of 1000 kg.

12. If the work done in moving a body is 1500 J and the time taken is 12 s, what is the mean value of the power?

13. A body having a mass of 300 kg is lifted through a distance of 200 m in 15 s. What is the average value of the power?

14. The output power of an electric motor is 8 kW and is maintained constant for 6 h. Calculate the work done (a) in kilowatt hours and (b) in megajoules.

15. If the speed of the motor in Q. 14 is 500 rev/min, what is the value of the torque in newton metres?

16. An engine, running at 1800 rev/min, develops a torque of 4 kN m. Calculate (a) the power developed and (b) the work done, in kilowatt hours, in 3 h.

17. A motor develops a torque of 3 kN m. At what speed must it run in order that it may develop 200 kW?

18. A locomotive, hauling a train at 110 km/h, exerts a pull of 10 kN. Calculate (a) the work done in 30 min (i) in megajoules, (ii) in kilowatt hours and (b) the power developed by the locomotive.

19. A bus having a mass of 3 Mg is travelling along a level road at 60 km/h. The resistance to motion is 10 N per kilonewton of weight. Calculate (a) the tractive effort required, (b) the power and (c) the work done in 40 min (i) in megajoules and (ii) in kilowatt hours.

20. A train has a mass of 150 Mg and is hauled at a constant speed of 90 km/h along a straight horizontal track. The track resistance is 6 mN per newton of weight. Calculate (a) the tractive effort in kilonewtons, (b) the work

done in 20 min (*i*) in megajoules and (*ii*) in kilowatt hours and (*c*) the power in kilowatts.

21. A barge is towed along a canal at 5 km/h. The tow rope is inclined at an angle of 30° to the direction of motion of the barge and the pull on the rope is 400 N. Calculate (*a*) the work done, in kilojoules, in 15 min and (*b*) the power required.

22. The rope used to haul a sledge along level ground is inclined at an angle 20° with the ground. The tension in the rope is 300 N. Calculate the work done when the sledge is hauled a distance of 80 m.

     If this work is done in 3 min, what is the average value of the power?

23. A belt friction brake applied at the circumference of a pulley, 400 mm diameter, exerts a backward drag of 220 N. If the speed of the pulley is 660 rev/min, calculate (*a*) the torque on the pulley and (*b*) the power absorbed by the brake.

     If the pulley is driven by an electric motor and if the input power to the motor is 3·6 kW, what is the efficiency of the motor?

24. In a brake test on an internal-combustion engine, the readings on the spring balances (fig. 8.7) were 490 N and 27 N. The effective diameter of the brake wheel was 0·9 m, and the speed of the engine was 500 rev/min. Calculate the output power of the engine.

25. A 150-mm diameter cylinder is being turned on a lathe driven at 180 rev/min. The tangential force exerted by the cutting tool is 800 N. If the efficiency of the lathe is 80 per cent, calculate the output power of the driving motor.

     If the corresponding efficiency of the motor is 75 per cent, calculate the input power to the motor.

26. A force of 2 kN acts on a plunger having a diameter of 200 mm. What is the average pressure on the plunger, in kilonewtons per square metre? If this plunger is used to force oil under a piston having a diameter of 500 mm, what is the mass of a body that the piston can support? What power is required to lift this mass through 100 mm in 4 s? Neglect all losses.

27. Calculate the kinetic energy, in kilojoules, possessed by a car having a mass of 900 kg when travelling at 100 km/h.

     What constant braking force would be required to bring the car to rest in 6 s?

28. Calculate the kinetic energy (*i*) in megajoules, (*ii*) in kilowatt hours, possessed by a body having a mass of 500 kg when travelling at 120 m/s.

     What constant force would be required to bring it to rest in 2 km?

29. A body having a mass of 50 kg is supported 40 m above ground. What is its potential energy?

     If the body is allowed to fall freely, calculate its potential and kinetic energies (*a*) when the body is 30 m above ground and (*b*) just before it touches the ground. Assume $g = 9·81$ m/s^2.

30. The energy expended in lifting a 30-kg mass is 6 kJ. Calculate the height through which it has been lifted.

     If this mass is then allowed to fall freely, what is its kinetic energy after it has fallen a distance of 10 m?

31. A locomotive, hauling a train at 90 km/h, exerts a pull of 20 kN. The total mass of the locomotive and carriages is 300 Mg. Calculate (*a*) the work done, in megajoules, in 10 min, (*b*) the power developed by the locomotive and (*c*) the kinetic energy of the locomotive and carriages, in megajoules (neglecting rotational inertia).

ANSWERS TO EXAMPLES 9

1. 7·5 N.
2. 314 kJ.
3. 120 m.
4. 15·3 kg, 150 W.
5. 388 N, 19·4 kJ.
6. 145·5 kJ, 40·4 W h; 7·275 kW.
7. 60 J.
8. 46·7 N; 1·633 J, 1·333 J.
9. 79·5 kJ.
10. 1·507 MJ, 2·354 MJ.
11. 222 MJ, 61·7 kW h; 15·4 h.
12. 125 W.
13. 39·24 kW.
14. 48 kW h, 172·8 MJ.
15. 153 N m.
16. 754 kW, 2262 kW h.
17. 637 rev/min.

18. 550 MJ, 153 kW h; 306 kW.
19. 294·3 N, 4·9 kW; 11·77 MJ, 3·27 kW h.
20. 8·83 kN; 265 MJ, 73·6 kW h; 221 kW.
21. 433 kJ, 482 W.
22. 22·56 kJ, 125·3 W.
23. 44 N m, 3·04 kW, 84·4 per cent.
24. 10·9 kW.
25. 1·413 kW, 1·884 kW.
26. 63·7 kN/m², 1275 kg, 312·5 W,
27. 347 kJ, 4·17 kN.
28. 3·6 MJ, 1 kW h, 1·8 kN.
29. 19·62 kJ; 14·715 kJ, 4·905 kJ; 0, 19·62 kJ.
30. 20·4 m, 2943 J.
31. 300 MJ, 500 kW, 93·75 MJ.

CHAPTER 10

Simple machines

10.1 Machines

A machine is a mechanical device for transmitting motion, force and energy. Lifting machines are usually arranged to enable a small *effort* (or driving force) to raise a much larger *load* (or resisting force). To achieve this, the effort must move through a much greater distance than that through which the load is raised, the work done by the effort being equal to the useful work in lifting the load together with the work required to overcome friction. This relationship is known as the *Principle of Conservation of Energy* and has already been referred to in section 9.10.

10.2 Mechanical advantage, velocity (or movement) ratio and efficiency

It was mentioned in section 10.1 that the purpose of most lifting machines is to enable a large load W to be moved by the application of a relatively small effort F. The ratio of the load to the effort is termed the *mechanical advantage* of the machine,

i.e. $$\text{mechanical advantage} = \frac{\text{load}}{\text{effort}} = \frac{W}{F} \qquad (10.1)$$

It is evident that for a given load W, the smaller the value of the effort F, the greater is the mechanical advantage.

The ratio of the distance moved by the effort to that moved by the load is termed the *movement ratio* or more commonly the *velocity ratio*,

i.e. $$\text{velocity ratio} = \frac{\text{distance moved by effort}}{\text{distance moved by load}} \qquad (10.2)$$

The value of the velocity ratio is dependent upon the arrangement of the machine and is *constant* for a given machine, whereas the value of the mechanical advantage, in general, decreases as the load is reduced, being zero at no load.

From the Principle of the Conservation of Energy,

work done by effort = work done on load
+ work done to overcome friction.

The work done in overcoming friction is converted into heat and is thus wasted as far as the machine is concerned. The ratio of the work done on the load to that done by the effort is termed the *efficiency* of the machine,

i.e. $$\text{efficiency} = \frac{\text{useful work done}}{\text{work done by effort}} \qquad (10.3)$$

Since work done by effort $F = F \times$ distance moved by effort and useful work done on load $W = W \times$ distance moved by load,

$$\therefore \quad \text{efficiency} = \frac{W \times \text{distance moved by load}}{F \times \text{distance moved by effort}}$$

$$= \frac{W}{F} \times \frac{\text{distance moved by load}}{\text{distance moved by effort}}$$

$$= \frac{\text{mechanical advantage}}{\text{velocity ratio}} \qquad (10.4)$$

For an *ideal* machine, i.e. a machine having no friction, the efficiency is unity ($= 100$ per cent), so that:

work done by effort = work done on load,

$\therefore$ *ideal* effort $\times$ distance moved by effort

$= W \times$ distance moved by load

so that $$\text{\textit{ideal} effort} = \frac{W}{\text{velocity ratio}} \qquad (10.5)$$

and *ideal* mechanical advantage $= \dfrac{W}{\text{ideal effort}}$

$= $ velocity ratio $\qquad (10.6)$

Let us now consider some of the simplest types of machines and determine the velocity ratio of each type.

10.3 The lever

The lever was probably the earliest device used by man to enable him to move a large load by means of the limited physical effort he could exert. For instance, guide books on Stonehenge have illus-

Engineering Science in SI Units

trations showing how men used wooden poles as levers to erect
the heaviest stones, each having a mass of about 50 Mg.

Fig. 10.1 shows a straight lever, pivotted at C. A mass, having
weight W, is suspended at B and a downward effort F is applied at A
to balance weight W.

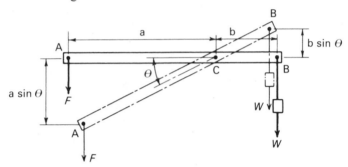

Fig. 10.1 A lever.

Suppose distances AC and BC to be a and b respectively, and the
weight of the lever to be negligible. If the lever is tilted anticlockwise
through an angle θ,

$$\left.\begin{array}{c}\text{distance through which effort } F\\ \text{moves in its own direction}\end{array}\right\} = a \sin \theta$$

so that   work done by effort $= F \times a \sin \theta$.

$$\left.\begin{array}{c}\text{Similarly, distance through which}\\ \text{load } W \text{ is moved in its own direction}\end{array}\right\} = b \sin \theta$$

so that   work done on load $= W \times b \sin \theta$

Hence,   velocity ratio $= \dfrac{a \sin \theta}{b \sin \theta} = \dfrac{a}{b}$

10.4 The wheel and axle

The wheel and axle may be regarded as an adaptation of the lever
to allow continuous rotaticn of the device about the pivot.

Fig.10.2 shows a wheel A, of radius a, and an axle B, of radius b,
carried by a shaft C. A body having weight W is attached to a cord
E, wound around axle B, and a downward effort F is applied to
cord D wound around the rim of wheel A.

When the wheel and axle make one revolution in an *anticlockwise*
direction, a length, $2\pi b$, of cord E is wound on axle B, thereby

lifting the load a distance $2\pi b$. At the same time, a length, $2\pi a$, of cord D is unwound from wheel A, enabling the effort to move a distance $2\pi a$.

Hence, $\qquad$ velocity ratio $= \dfrac{\text{distance moved by effort}}{\text{distance moved by load}}$

$$= \frac{2\pi a}{2\pi b} = \frac{a}{b}$$

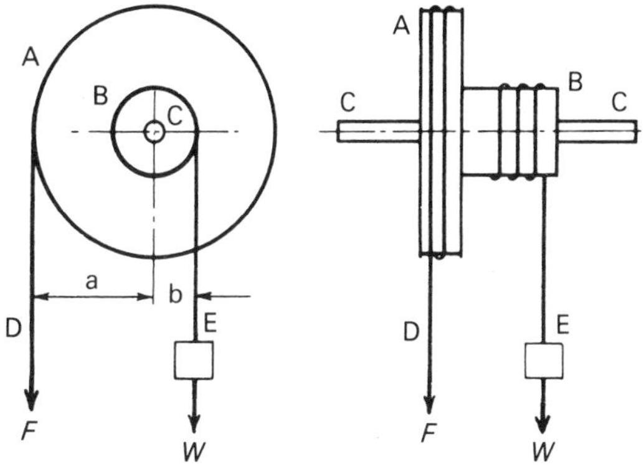

Fig. 10.2 The wheel and axle.

Example 10.1 *In a certain wheel and axle machine, the diameters of the wheel and axle are* 450 *mm and* 60 *mm respectively. The efficiency is* 97 *per cent (or* 0·97 *per unit) when a body having a mass of* 40 *kg is being lifted. Calculate* (a) *the velocity ratio,* (b) *the ideal effort,* (c) *the actual effort and* (d) *the mechanical advantage.*

$$\text{Weight of body lifted} = W \simeq 40 \times 9\cdot81$$
$$= 392\cdot4 \text{ N.}$$

(a) $\qquad$ Velocity ratio $= \dfrac{\text{diameter of wheel}}{\text{diameter of axle}}$

$$= \frac{450 \text{ [mm]}}{60 \text{ [mm]}} = 7\cdot5.$$

(b) From expression (10.5), we have:

$$\text{ideal effort} = \frac{W}{\text{velocity ratio}}$$

$$= \frac{392\cdot4 \text{ [N]}}{7\cdot5} = 52\cdot3 \text{ N.}$$

(*c*) From expression (10.4), we have:

$$\text{efficiency} = \frac{\text{load}}{\text{effort}} \times \frac{1}{\text{velocity ratio}}$$

$$\therefore \qquad 0.97 = \frac{392.4 \text{ [N]}}{F} \times \frac{1}{7.5}$$

so that actual effort $= F = 54 \text{ N}$.

Alternatively,   actual effort $= \dfrac{\text{ideal effort}}{\text{efficiency}}$

$$= 52.3 \text{ [N]}/0.97 = 54 \text{ N}.$$

(*d*)   Mechanical advantage $= \dfrac{W}{F} = \dfrac{392.4 \text{ [N]}}{54 \text{ [N]}} = 7.275.$

Alternatively, from expression (10.4), we have:

$$\text{mechanical advantage} = \text{velocity ratio} \times \text{efficiency}$$
$$= 7.5 \times 0.97 = 7.275.$$

10.5 Inclined plane

As mentioned in section 4.1, the principle of the inclined plane was known at least 5000 years ago when it was employed in the construction of the Pyramids. The smooth inclined plane was discussed in Chapter 4, but in this section we shall consider the general case where friction may not be negligible.

Suppose a body having weight W to be hauled up the whole of the inclined surface (fig. 10.3) by a force F acting parallel to the plane.

Hence,   work done by effort $= F \times$ AB.

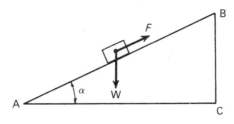

Fig. 10.3 The inclined plane.

While the load has been hauled a distance AB up the inclined plane, it has also been lifted through the vertical distance BC, so that work done on load $= W \times$ BC.

$$\text{Velocity ratio} = \frac{\text{AB}}{\text{BC}} = \frac{1}{\sin \alpha}$$

where α is the angle between the plane and the horizontal.

10.6 The screw and the screw-jack

The screw is really an inclined plane converted into a helical inclined path around the bolt on which the screw has been cut, thereby enabling the circular motion of a nut to be converted into a linear motion along the screwed portion of the bolt. This means that a small effort applied tangentially at the end of, say, a spanner moves through a distance equal to ($2\pi \times$ radius of effort) while the nut travels a distance equal to the lead of the thread, i.e. the distance between the centres of adjacent threads for single-start thread.

Hence, $\qquad$ velocity ratio $= \dfrac{2\pi \times \text{radius of effort}}{\text{lead of thread}}$.

e.g. if the radius at which the effort is applied is, say, 200 mm and the lead of the thread is 1·8 mm,

$$\text{velocity ratio} = \frac{2\pi \times 200 \text{ [mm]}}{1 \cdot 8 \text{ [mm]}} = 698.$$

Allowing for an efficiency of, say, 15 per cent, we have from expression (10.4):

$$\text{mechanical advantage} = \text{velocity ratio} \times \text{efficiency}$$

$$= 698 \times 0 \cdot 15 = 105.$$

A device, using the principle of the screw, that is commonly employed for lifting heavy objects is the screw-jack shown in fig. 10.4.

An effort F is applied tangentially at the end of an arm of radius r. For a screw having a right-hand thread, one turn of the effort in an anticlockwise direction, viewed from above the jack, lifts the load W through a distance equal to the *lead*, l, of the screw. (In a single-start screw, the lead is equal to the pitch of the thread.)

For one revolution of the effort,

$$\text{distance moved by effort} = 2\pi r$$

and $\quad$ distance moved by load $= l$

$\therefore \qquad$ velocity ratio $= 2\pi r/l$.

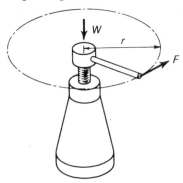

Fig. 10.4 The screw-jack.

Example 10.2 *A body having a mass of 200 kg is resting on top of a screw-jack. The screw has a lead of 8 mm, and an effort of 36 N has to be applied tangentially at a radius of 250 mm to lift the load. Calculate* (a) *the velocity ratio,* (b) *the mechanical advantage and* (c) *the efficiency.*

Weight of body supported on screw-jack $\simeq$ 200 × 9·81
$$= 1962 \text{ N.}$$

(a)   Velocity ratio $= 2\pi r/l$
$$= 2\pi \times 250 \text{ [mm]}/8 \text{ [mm]} = 196.$$

(b)   Mechanical advantage $= W/F$
$$= 1962 \text{ [N]}/36 \text{ [N]} = 54·5.$$

(c)   Efficiency $= \dfrac{\text{mechanical advantage}}{\text{velocity ratio}}$
$$= 54·5/196 = 0·278 \text{ per unit}$$
$$= 27·8 \text{ per cent.}$$

10.7 Pulleys

A single-pulley system. Fig. 10.5(a) shows a pulley block fitted with one pulley (or sheave), with a rope passing over the pulley and supporting at one end a body having weight W. This load can be lifted by applying an effort F to the other end of the rope. With this simple arrangement, the velocity ratio is unity since the distance through which the effort is applied is exactly the same as that through which the load is raised.

F

The effort F has to be greater than the load W to allow for friction at the pulley, so that the mechanical advantage ($= W/F$) is less than unity. The only advantage of the single-pulley system is that a person, when hauling a rope downwards, is able to make use of his own weight when exerting the pull and thus finds it easier than to lift the load directly.

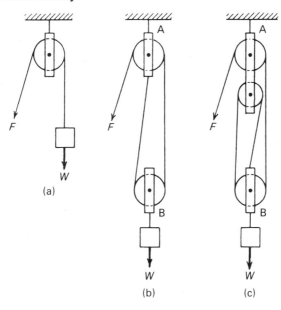

Fig. 10.5 Arrangements of pulley-blocks.

A two-pulley system. The velocity ratio can be increased by using more pulleys; e.g. in fig. 10.5(b) there are two pulley-blocks A and B, each with one pulley. A rope has one end attached to pulley-block A and passes round the pulleys of B and A, as shown. The body to be lifted is attached to pulley-block B.

If the load W were raised 1 m by means of effort P, then *each* of the lengths of rope between the pulley-blocks would be shortened by 1 m and the effort F would move 2 m. Hence,

$$\text{velocity ratio} = \frac{\text{distance moved by effort}}{\text{distance moved by load}}$$

$$= \frac{2\,[\text{m}]}{1\,[\text{m}]} = 2.$$

A three-pulley system. Figure 10.5(c) shows pulley-block A fitted with two pulleys. These pulleys are usually on the same spindle, but for convenience of explanation, they are shown one above the other. With this arrangement, effort F moves a distance of 3 m when the load is lifted 1 m, so that the velocity ratio is now 3.

In general, if n be the *total* number of pulleys (or sheaves) on the two pulley-blocks,

$$\text{velocity ratio} = n.$$

Example 10.3 *A body having a mass of 30 kg was lifted by an effort of 95 N by means of a pulley-block arrangement. The upper and lower blocks had 3 pulleys and 2 pulleys respectively. Calculate* (a) *the velocity ratio,* (b) *the mechanical advantage and* (c) *the efficiency.*

(a) Total number of pulleys $= 3 + 2 = 5$,

$\therefore$ velocity ratio $= 5$.

(b) Weight of body $\simeq 30 \times 9\cdot81 = 294\cdot3$ N,

$\therefore$ mechanical advantage $= \dfrac{W}{F} = \dfrac{294\cdot3\,[\text{N}]}{95\,[\text{N}]} = 3\cdot1$.

(c) Efficiency $= \dfrac{\text{mechanical advantage}}{\text{velocity ratio}}$

$= 3\cdot1/5 = 0\cdot62$ per unit

$= 62$ per cent.

10.8 Belt and chain drives

A belt or a chain is used when a shaft has to be driven from a parallel shaft that is too far away for the use of gear wheels (section 10.9). Fig. 10.6 shows a belt drive in which A is the *driver* pulley and B the *driven* pulley. The transfer of motion from pulley A to the belt and again from the belt to pulley B is dependent upon friction at each area of contact between belt and pulley.

If d_A and d_B = diameters of pulleys A and B respectively and n_A and n_B = speeds, in revolutions/second, of A and B respectively,

then   linear speed of rim of pulley A $= \pi d_A n_A$

and   linear speed of rim of pulley B $= \pi d_B n_B$.

If there is no slipping, the linear speed of the rim of each pulley is the same as the speed of the belt, hence $$\pi d_A n_A = \pi d_B n_B$$

$$\therefore \quad \frac{\text{speed of driver pulley A}}{\text{speed of driven pulley B}} = \frac{\text{diameter of driven pulley B}}{\text{diameter of driver pulley A}} \quad (9.7)$$

i.e. the speeds of the pulleys are inversely proportional to their diameters.

When a driving torque is applied to driver pulley A, the side of the belt approaching the pulley tightens and that leaving the pulley slackens. Thus, with the driver pulley rotating clockwise as in fig. 10.6, the lower side of the belt has a larger tension than the upper side. side.

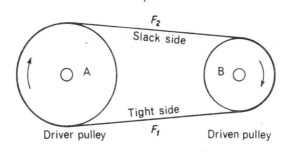

Fig. 10.6 A belt drive.

If F_1 and F_2 be the tensions in the tight and slack sides respectively of the belt,

effective force due to friction $= F_1 - F_2$

and power transmitted $=$ net force [newtons]

$\times$ speed of belt [metres/second]

$$= (F_1 - F_2) \times \pi d_A n_A \text{ watts} \quad (10.8)$$
$$\text{or} = (F_1 - F_2) \times \pi d_B n_B \text{ watts} \quad (10.9)$$

These expressions are similar to expression (9.6) deduced for a brake test in section 9.6.

One of the main disadvantages of a belt is its liability to slip. This disadvantage can be avoided by the use of a chain whose links engage with teeth on the driver and driven wheels. One of the best-known examples is the use of a chain to transmit power from the pedals to the rear wheel of a bicycle.

Since the number of teeth on each wheel is proportional to the diameter of the wheel,

$$\frac{\text{speed of driver wheel}}{\text{speed of driven wheel}} = \frac{\text{diameter of driven wheel}}{\text{diameter of driver wheel}}$$

$$= \frac{\text{number of teeth on driven wheel}}{\text{number of teeth on driver wheel}}.$$

Example 10.4 *A belt-driven pulley has a diameter of* 500 *mm and its speed is* 300 *rev/min. The tensions in the two sides of the belt are* 1800 *N and* 400 *N respectively. Calculate the power transmitted by the belt.*

Effective friction force $= 1800 - 400 = 1400$ N.

Linear speed of belt $= \pi \times 0.5$ [m] $\times$ (300/60) [rev/s]
$= 7.85$ m/s,

$\therefore$ power transmitted $= 1400$ [N] $\times 7.85$ [m/s]
$= 11\,000$ W $= 11$ kW.

Alternatively,   speed $= 300/60 = 5$ rev/s
and torque $= 1400$ [N] $\times 0.25$ [m] $= 350$ N m,

hence, from expression (9.4), we have:

power transmitted $= 2\pi n T$
$= 2\pi \times 5$ [rev/s] $\times 350$ [N m]
$= 11\,000$ W $= 11$ kW.

10.9 Gear wheels

Gear wheels are used to transmit motion and power from one shaft to a parallel shaft in close proximity, and to enable the speed of rotation to be stepped up or down to suit specific requirements, e.g. to reduce the high speed of an electric motor to the relatively low speed of a lathe.

A gear wheel has a number of specially-shaped teeth around its periphery. These teeth mesh with similar teeth on a second wheel as shown in fig. 10.7, where wheel A (the *driver*) drives wheel B (the *follower*).

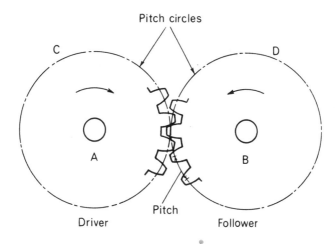

Fig. 10.7 Gear wheels.

In order that the teeth may mesh correctly, the pitch of the teeth, i.e. the peripheral distance between the centres of adjacent teeth, measured on the pitch circles C and D, must be the same for the two wheels. Hence, for a given pitch, the number of teeth is proportional to the diameter of the pitch circle;

$$\therefore \quad \frac{\text{number of teeth on wheel A}}{\text{number of teeth on wheel B}} = \frac{\text{circumference of wheel A}}{\text{circumference of wheel B}}.$$

If n_A and n_B be the speeds of wheels A and B respectively, in revolutions per second,

peripheral speed of driver A, measured at pitch circle
= circumference of A $\times\ n_A$

and peripheral speed of follower B, measured at pitch circle
= circumference of B $\times\ n_B$.

Since there can be no slip, the peripheral speed at the pitch circles is the same for the two wheels,

i.e. circumference of A $\times\ n_A$ = circumference of B $\times\ n_B$

$$\therefore \quad \frac{\text{speed of driver A}}{\text{speed of follower B}} = \frac{\text{circumference of follower}}{\text{circumference of driver}}$$

$$= \frac{\text{number of teeth on follower}}{\text{number of teeth on driver}} \quad (10.10)$$

i.e. the speeds of the gear wheels are inversely proportional to the number of teeth on the wheels.

It will be noted that gear wheels, A and B, in fig. 10.7 rotate in opposite directions. The directions of rotation of A and B can be arranged to be the same by introducing an idle wheel C between A and B as shown in fig. 10.8. If idler C rotates at n_C revolutions per second,

$$\frac{n_A}{n_C} = \frac{\text{number of teeth on C}}{\text{number of teeth on A}}.$$

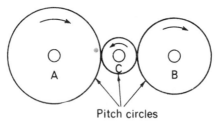

Pitch circles

Fig. 10.8 Effect of an idle wheel.

Similarly, $$\frac{n_C}{n_B} = \frac{\text{number of teeth on B}}{\text{number of teeth on C}}$$

$$\therefore \quad \frac{n_A}{n_C} \times \frac{n_C}{n_B} = \frac{\text{number of teeth on C}}{\text{number of teeth on A}} \times \frac{\text{number of teeth on B}}{\text{number of teeth on C}}$$

hence, $$\frac{n_A}{n_B} = \frac{\text{number of teeth on B}}{\text{number of teeth on A}}$$

i.e. the gear ratio of the gear train ABC is independent of the number of teeth on the idler.

Example 10.5 *In the compound gear train shown in fig. 10.9, gear wheels B and C are rigidly attached to an idle shaft. Wheels A, B and C have 60, 20 and 50 teeth respectively, and the speed of A is 200 rev/min. Calculate the number of teeth on wheel D so that its speed may be 1000 rev/min.*

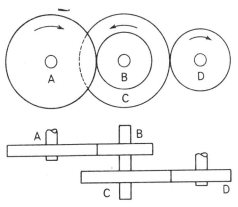

Fig. 10.9 A compound gear train.

$$\frac{\text{Speed of B}}{\text{Speed of A}} = \frac{\text{number of teeth on A}}{\text{number of teeth on B}} = \frac{60}{20} = 3.$$

∴   speed of B = 3 × 200 [rev/min] = 600 rev/min.

Since gear wheels B and C are fixed to the same shaft, their speeds must be the same,

∴   speed of C = 600 rev/min.

Also, $\dfrac{\text{speed of C}}{\text{speed of D}} = \dfrac{\text{number of teeth on D}}{\text{number of teeth on C}}$

i.e. $\dfrac{600}{1000} = \dfrac{\text{number of teeth on D}}{50}$

∴   number of teeth on D = 30.

Summary of Chapter 10

Several simple machines have been discussed which enable a small effort to lift a relatively large load. Also, transmission of motion and power by belt, chain and gear wheels have been described.

By the Principle of Conservation of Energy,

$$\text{work done by effort} = \text{work done on load} + \text{work done in overcoming friction}$$

$$\text{Mechanical advantage} = \frac{\text{load}}{\text{effort}} \tag{10.1}$$

$$\text{Velocity ratio} = \frac{\text{distance moved by effort}}{\text{distance moved by load}} \quad (10.2)$$

$$\text{Efficiency} = \frac{\text{work done on load}}{\text{work done by effort}} \quad (10.3)$$

$$= \frac{\text{mechanical advantage}}{\text{velocity ratio}} \quad (10.4)$$

$$\text{Ideal effort} = \frac{\text{load}}{\text{velocity ratio}} \quad (10.5)$$

$$\text{Ideal mechanical advantage} = \frac{\text{load}}{\text{ideal effort}}$$

$$= \text{velocity ratio} \quad (10.6)$$

For belt drive,

$$\frac{\text{speed of driver pulley}}{\text{speed of driven pulley}} = \frac{\text{diameter of driven pulley}}{\text{diameter of driver pulley}} \quad (10.7)$$

and   power transmitted $= (F_1 - F_2) \times \pi dn \quad (10.8)$

For gear wheels,

$$\frac{\text{speed of driver}}{\text{speed of follower}} = \frac{\text{number of teeth on follower}}{\text{number of teeth on driver}} \quad (10.10)$$

EXAMPLES 10

1. In a certain lifting machine, the effort moves 0·3 m for every millimetre lift of the load. If the efficiency of the machine is 45 per cent when the load is 6 kN, calculate (a) the velocity ratio, (b) the effort required and (c) the mechanical advantage.

2. In a certain machine, an effort of 40 N was required to raise a body having a mass of 50 kg. When the body was raised a distance of 30 mm, the effort moved through a distance of 540 mm. Calculate (a) the velocity ratio, (b) the mechanical advantage, (c) the efficiency, (d) the ideal effort and (e) the ideal mechanical advantage.

3. A lever AB, 3 m long, is pivotted at a point 0·4 m from end A. Calculate the value of the load, in kilonewtons, at end A that can be raised by a downward effort of 500 N applied at right angles to the lever at end B. Assume the weight of the lever and the friction at the pivot to be negligible. Also calculate the mechanical advantage of the arrangement.

4. A crowbar, 2 m long, is used to move an object at one end of the bar. If the force required to move the object is 800 N and the maximum effort that can be exerted at the other end of the lever is 120 N, calculate the maximum distance of the fulcrum from the end of the bar at which the load is located.

5. The diameters of the wheel and axle (fig. 9.2) are 300 mm and 75 mm respectively. When a body having a weight of 300 N is being lifted, the efficiency is 80 per cent. Calculate (*a*) the velocity ratio, (*b*) the effort required, (*c*) the mechanical advantage, (*d*) the useful work done when the load is raised 20 mm and (*e*) the corresponding work done by the effort.

6. In a wheel-and-axle machine, the wheel has a diameter of 200 mm and the axle has a diameter of 40 mm. Calculate (*a*) the effort required to lift a mass of 15 kg, if the efficiency is 0·9 per unit, (*b*) the velocity ratio, (*c*) the mechanical advantage, (*d*) the ideal effort, (*e*) the effort to overcome friction and (*f*) the ideal mechanical advantage.

7. A simple winch has a drum, 250 mm diameter, around which a rope is wound. The handle, which operates at a radius of 400 mm, is connected directly to the drum. The winch is used to raise a body having a mass of 50 kg. Calculate (*a*) the velocity ratio, (*b*) the tangential effort required at the handle if the efficiency of the winch is 80 per cent.

8. A block of concrete having a mass of 200 kg is hauled up an inclined plane having a gradient of 1 in 5. If the efficiency is 35 per cent, calculate (*a*) the effort required if its direction is parallel to the plane, (*b*) the velocity ratio and (*c*) the mechanical advantage.

9. A body having a weight of 3 kN is dragged up a surface inclined at an angle of 20° to the horizontal. The effort, acting parallel to the surface, is 1·3 kN. Calculate (*a*) the velocity ratio, (*b*) the mechanical advantage, (*c*) the efficiency, (*d*) the work done in hauling the body a distance of 3 m and (*e*) the corresponding useful work done.

10. The screw of a certain screw-jack has a lead of 10 mm. An effort of 120 N is applied tangentially at the end of a bar having a radius of 250 mm. If the efficiency is 0·3 per unit, calculate (*a*) the load, in kilonewtons, (*b*) the velocity ratio and (*c*) the mechanical advantage.

11. A screw-jack is used to lift a mass of 500 kg. The lead of the screw is 12 mm. Calculate the force required at the end of a 300-mm arm, assuming the efficiency to be 28 per cent. Also determine the velocity ratio and the mechanical advantage.

12. A lifting tackle consists of two pulley blocks. Each block has two pulleys (or sheaths). One end of the cord is attached to the upper pulley block. Sketch the arrangement and calculate (*a*) the velocity ratio, (*b*) the effort required to lift a mass of 120 kg, assuming the efficiency to be 85 per cent and (*c*) the mechanical advantage.

13. A system of pulleys consists of an upper block fitted with three pulleys and a lower block fitted with two pulleys. An effort of 240 N is required to raise a body having a mass of 85 kg. Sketch the arrangement and calculate (*a*) the velocity ratio, (*b*) the mechanical advantage, (*c*) the efficiency, (*d*) the work done in lifting the body 50 mm and (*e*) the corresponding work done by the effort.

14. A pulley, 150 mm in diameter and rotating at 900 rev/min, is belt-coupled to another pulley. What should be the diameter of the latter pulley in order that it may rotate at 280 rev/min?

15. A shaft A has a pulley 120 mm in diameter from which shaft B is to be driven by a belt. Shaft B has to drive a shaft C, the pulley on the latter

being 400 mm diameter. The speeds of shafts A, B and C are to be 900, 500 and 180 rev/min respectively. Calculate the diameters of the two pulleys on shaft B.

16. Energy is transmitted from the shaft of an engine by a belt passing over a pulley having a diameter of 600 mm. The effective pull on the belt is 2·8 kN. Calculate (a) the torque, in newton metres, and (b) the power transmitted when the speed is 4 rev/s.

17. The pull on the tight side of a belt is 720 N and that on the slack side is 80 N. The driving pulley has a diameter of 460 mm and is rotating at 480 rev/min. Calculate (a) the torque and (b) the power transmitted.

18. A motor drives a line of shafting by means of a chain drive. The wheel on the motor has 23 teeth and that on the shafting 82 teeth. If the speed of the motor is 900 rev/min, what is the speed of the shafting?

19. By applying a force of 200 N to the pedal of his bicycle, a cyclist exerts a turning moment of 35 N m on the crank. The bicycle has 48 teeth on the front chain-wheel and 18 on the rear sprocket-wheel. The wheels are 700 mm in diameter. Calculate (a) the effective length of the crank, (b) the speed of the rear wheel, in revolutions/minute, when the pedals are being driven at 180 rev/min, (c) the velocity ratio of the machine, i.e. the ratio of the linear speed of the pedals to the linear speed of the outermost surface of the rear wheel, and (d) the force, parallel to the road, acting on the machine. Neglect any losses.

20. Two gear wheels, A and B, have 80 and 30 teeth respectively. If the speed of A is 600 rev/min, what is the speed of B?

21. Three gear wheels, A, B and C, are on three parallel shafts. A has 14 teeth and meshes with B, which has 37 teeth and meshes with C which has 49 teeth. If the speed of A is 1300 rev/min, what is the speed of C?

22. In a double-reduction gear train (fig. 10.9), wheel A has 24 teeth and rotates at 80 rev/min. Wheels B and C have 96 and 36 teeth respectively and are rigidly attached to each other. If wheel D has 60 teeth, calculate its speed.

ANSWERS TO EXAMPLES 10

1. 300, 44·4 N, 135.
2. 490·5 N, 18, 12·26, 68·1 per cent, 27·25 N, 18.
3. 3·25 kN, 6·5.
4. 0·261 m.
5. 4, 93·75 N, 3·2, 6 J, 7·5 J.
6. 32·7 N, 5, 4·5, 29·43 N, 3·27 N, 5.
7. 3·2, 192 N.
8. 1120 N, 5, 1·75.
9. 2·92, 2·31, 79 per cent, 3900 J, 3080 J.
10. 5·65 kN, 157, 47·1.
11. 111·8 N, 157, 44.

12. 4, 346 N, 3·4.
13. 5, 3·48, 69·6 per cent, 41·7 J, 60 J.
14. 482 mm.
15. 216 mm, 144 mm.
16. 840 N m, 21·1 kW.
17. 147·2 N m, 7·4 kW.
18. 252·5 rev/min.
19. 175 mm, 480 rev/min, 0·1875, 37·5 N.
20. 1600 rev/min.
21. 371·4 rev/min.
22. 120 rev/min.

CHAPTER 11

Stress and strain

11.1 Stress

When a material has a force exerted on it, the material is said to be *stressed* or in a state of *stress*. If a rod is subjected to a *tension*, the force per unit area of cross-section of the rod is referred to as *tensile stress*. If the rod is subjected to compression, the force per unit area of cross-section is termed a *compressive stress*. Hence, in general:

$$\text{stress} = \frac{\text{force}}{\text{cross-sectional area}} \qquad (11.1)$$

Tensile and compressive stresses are sometimes referred to as *normal stresses* since they act at right angles to the cross-sectional area which is used for calculating the value of the stress.

The *SI unit of stress* is the *newton per square metre* (symbol, N/m^2). At the 1971 meeting of the CGPM, it was decided to adopt the term *pascal* (symbol, Pa) for this unit, in memory of the French philosopher Blaise Pascal (1623-62) who carried out many brilliant experiments in hydrostatics and pneumatics.

It is often more convenient to express stress in kilonewtons per square metre (kN/m^2) or kilopascals (kPa), or in meganewtons per square metre (MN/m^2) or megapascals (MPa) where

$$1 \text{ kN/m}^2 \text{ or } 1 \text{ kPa} = 10^3 \text{ N/m}^2 \text{ or } 10^3 \text{ Pa}$$

and $\qquad 1 \text{ MN/m}^2 \text{ or } 1 \text{ MPa} = 10^6 \text{ N/m}^2 \text{ or } 10^6 \text{ Pa.}$

Example 11.1 *A tie-bar has a cross-sectional area of* 125 mm² *and is subjected to a pull of* 10 kN. *Calculate the stress in meganewtons per square metre or megapascals.*

$$\text{Force} = 10 \text{ kN} = 10\,000 \text{ N.}$$

$$\text{Cross-sectional area} = 125 \text{ mm}^2 = 125 \times 10^{-6} \text{ m}^2,$$

$$\therefore \qquad \text{stress} = \frac{10\,000 \text{ [N]}}{125 \times 10^{-6} \text{ [m}^2]} = 80 \times 10^6 \text{ N/m}^2$$

$$= 80 \text{ MN/m}^2 \text{ or MPa.}$$

Example 11.2 *A wire, 1·5 mm diameter, supports a mass of 60 kg. Calculate the stress.*

Tension in wire = weight of the mass

$$\simeq 60 \times 9{\cdot}81 = 588{\cdot}6 \text{ N.}$$

Cross-sectional area $= (\pi/4) \times (1{\cdot}5)^2 = 1{\cdot}767 \text{ mm}^2$

$$= 1{\cdot}767 \times 10^{-6} \text{ m}^2$$

$$\therefore \quad \text{stress} = \frac{588{\cdot}6 \text{ [N]}}{1{\cdot}767 \times 10^{-6} \text{ [m}^2\text{]}} = 333{\cdot}5 \times 10^6 \text{ N/m}^2$$

$$= 333{\cdot}5 \text{ MN/m}^2 \text{ or MPa.}$$

11.2 Strain

When a rod or a wire is pulled, it stretches; and the total stretch or elongation, expressed as a fraction of the unstretched length, is termed *direct strain* or merely *strain* when it is obvious that the change in length is due to tension or compression. Thus, if the pull on a rod having an unstretched length of 1 m produces an extension of 2·5 mm,

$$\text{strain} = \frac{\text{extension}}{\text{original length}} \qquad (11.2)$$

$$= \frac{2{\cdot}5 \text{ [mm]}}{(1 \times 1000) \text{ [mm]}} = 0{\cdot}0025.$$

It will be noticed that the extension and the original length must be expressed in the same unit. In the above expression, they are both expressed in millimetres, but the result would be exactly the same if they were both expressed in, say, metres, thus:

$$\text{strain} = \frac{(2{\cdot}5/1000) \text{ [m]}}{1{\cdot}0 \text{ [m]}} = 0{\cdot}0025.$$

Hence strain is merely a ratio and possesses no units.

11.3 Tensile test on a steel wire

The behaviour of various materials, when stretched by gradually increasing forces, may be studied experimentally in the laboratory.

The results of a test performed on a steel wire are given in the following table. The initial length of the wire was 907 mm and the initial cross-sectional area was 0·34 mm².

Load [N]	0	10	20	30	40	50	60	70	80	90	100
Extension [mm]	0	0·10	0·22	0·35	0·50	0·62	0·75	1·1	2·1	4·0	7·5

Load [N]	105	110	118	129	136	145	150	154	
Extension [mm]	10·4	12·9	17·8	21·8	29·1	42·7	54	70	(broke)

An examination of the tabulated figures shows that there are two stages in the stretching of the wire:

(*a*) *The Elastic Stage.* Up to a load of about 60 N, the extension is small and is proportional to the load, as shown in fig. 11.1. During this stage of the test, it is found that if the load is removed, the stretch disappears and the wire returns to its original length. The fact that during this elastic stage the extension is proportional to the load is characteristic of the elastic straining of materials.

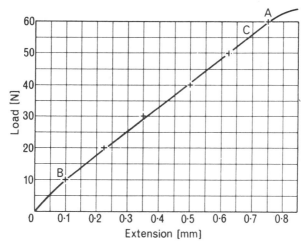

Fig. 11.1 Elastic extension of a wire.

(*b*) *The Plastic Stage.* The results for loads of 80 N upwards are plotted in fig. 11.2, where the extension is shown to a much smaller scale than in fig. 11.1.

It is evident from the table and from fig. 11.2 that for loads exceeding about 60 N, the extension for a given increase in load is much greater than for loads below 60 N. For instance, when the load is increased from 60 N to 150 N, the extension increases from 0·75 mm to 54 mm, i.e. when the load is increased about 2·5 times, the extension increases more than 70 times.

During this plastic stage of the test, it is found that when the load is increased from 60 N to 150 N, the extension increases from

is said to have a *plastic deformation* or *permanent set*. A metal that is able to undergo cold plastic deformation, usually as the result of tension, is said to be *ductile* or to possess *ductility*.

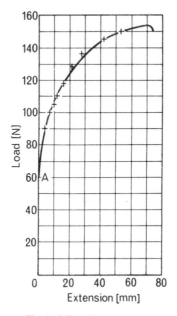

Fig. 11.2 Tensile test on a steel wire.

11.4 Hooke's law

Robert Hooke (1635–1703), a famous English physicist, architect and inventor, carried out a series of experiments with springs and wires and was the first to publish, in 1676, a clear statement to the effect that when a material is worked within its elastic range, *the extension is proportional to the force.*

11.5 Modulus of elasticity (direct) or Young's modulus

For a material worked within its elastic range, we have from Hooke's law that since the extension is proportional to the force:

$$\text{strain} \propto \text{stress},$$

i.e.

$$\frac{\text{stress}}{\text{strain}} = \text{a constant.}$$

This constant is termed *modulus of elasticity (direct)* or *Young's modulus*, after Thomas Young* who was the first to determine this constant for some materials.

The symbol for modulus of elasticity (direct) or Young's modulus is E,

i.e.
$$E = \frac{\text{tensile or compressive stress}}{\text{strain}} \qquad (11.3)$$
$$= \frac{\text{force/original cross-sectional area}}{\text{change of length/original length}}$$

Young's modulus may be regarded as a measure of the resistance which a material offers to extension or compression, i.e. the larger the value of E, the smaller is the extension or compression produced by a given stress.

When the value of Young's modulus is being determined from experimental results such as those shown in fig. 11.1, it is usually better not to use figures for load and extension at any one point on the graph but to take the *difference* in extension and the corresponding *difference* in load between two points, such as B and C in fig. 11.1, on a straight line drawn through the points determined experimentally. This is due to the difficulty in determining accurately the point representing zero load and zero extension.

Let us now proceed to calculate the value of E from the graph of fig. 11.1:

$$\text{load at B} = 10 \text{ N,}$$
$$\text{load at C} = 55 \text{ N,}$$
$$\therefore \quad \text{increase in load} = 45 \text{ N.}$$

$$\text{Initial cross-sectional area} = 0.34 \text{ mm}^2 \text{ (see section 11.3)}$$
$$= 0.34 \times 10^{-6} \text{ m}^2,$$
$$\therefore \quad \text{increase in stress} = 45 \text{ [N]}/(0.34 \times 10^{-6}) \text{ [m}^2\text{]}$$
$$= 132.4 \times 10^6 \text{ N/m}^2 \text{ or Pa.}$$

$$\text{Extension at B} = 0.10 \text{ mm}$$

* Thomas Young (1773-1829) was one of the most versatile geniuses in history. He studied languages and medicine and was a Professor of Physics at the Royal Institution, London. He was the first to translate the Egyptian hieroglyphics on the Rosetta Stone at the British Museum and was also famous for his share in establishing the undulatory theory of light.

and   extension at C = 0·69 mm,

∴   increase in extension = 0·59 mm.

Initial length of wire = 907 mm (see section 11·3)

∴   increase in strain = 0·59 [mm]/907 [mm]

= 0·65 × 10⁻³.

Hence $E = \dfrac{\text{increase in stress}}{\text{increase in strain}} = \dfrac{132\cdot4 \times 10^6 \text{ [N/m}^2]}{0\cdot65 \times 10^{-3}}$

$= 204 \times 10^9$ N/m² or Pa

$= 204\,000$ MN/m² or MPa

$= 204$ giganewtons/metre²

(or GN/m² or GPa).

Example 11.4 *A mild-steel rod, 4 m long and 30 mm in diameter, carries a tensile load of 100 kN. Calculate the extension, assuming E = 200 GPa.*

Cross-sectional area = $(\pi/4) \times (30)^2 = 706\cdot5$ mm²

$= 706\cdot5 \times 10^{-6}$ m²,

tensile load = 100 kN = 100 000 N,

∴   stress = 100 000 [N]/(706·5 × 10⁻⁶) [m²]

$= 1\cdot415 \times 10^8$ Pa.

$E = 200$ GPa $= 200 \times 10^9$ Pa.

Substituting in expression (11.3), we have:

200×10^9 [Pa] $= \dfrac{1\cdot415 \times 10^8 \text{ [Pa]}}{\text{strain}}$

∴   strain = $7\cdot075 \times 10^{-4}$

and   extension = $7\cdot075 \times 10^{-4} \times 4$ m $= 2\cdot83 \times 10^{-3}$ m

= 2·83 mm.

Example 11.5 *A member of a structure is 5 m long and has a cross-sectional area of 900 mm². What is the greatest pull which can be applied if the elongation is not to exceed 2·8 mm? Assume E = 220 GPa.*

Maximum allowable strain = $\dfrac{2\cdot8 \text{ [mm]}}{(5 \times 1000) \text{ [mm]}}$

$= 0\cdot56 \times 10^{-3}$

and $E = 220$ GPa $= 220 \times 10^9$ Pa.

Substituting in expression (11.3), we have:

$$220 \times 10^9 \text{ [Pa]} = \frac{\text{maximum allowable stress}}{0 \cdot 56 \times 10^{-3}}$$

$\therefore$ maximum allowable stress $= 123 \cdot 2 \times 10^6$ Pa.

Cross-sectional area $= 900$ mm^2 $= 900 \times 10^{-6}$ m^2,

$\therefore$ maximum allowable pull $= (123 \cdot 2 \times 10^6)$ [Pa]

$\times$ (900×10^{-6}) [m^2]

$= 110\ 900$ N $= 110 \cdot 9$ kN.

Example 11.6 *An extensometer used on a tie-bar of a steel bridge shows an extension of $0 \cdot 11$ mm on a length of 200 mm. Assuming $E = 210$ GPa, calculate the stress in the tie-bar.*

$$\text{Strain} = \frac{0 \cdot 11 \text{ [mm]}}{200 \text{ [mm]}} = 0 \cdot 55 \times 10^{-3}$$

and $E = 210$ GPa $= 210 \times 10^9$ Pa.

Substituting in expression (11.3), we have:

$$210 \times 10^9 \text{ [Pa]} = \frac{\text{stress}}{0 \cdot 55 \times 10^{-3}}$$

$\therefore$ stress $= 115 \cdot 5 \times 10^6$ Pa

$= 115 \cdot 5$ MPa.

11.6 Tensile-testing of materials

In a tensile test measurements are made of the force used to extend a standard test piece at a constant rate, with the resulting elongation of a specified gauge length of the test piece being measured by some form of extensometer.

In order to eliminate any variations in tensile test data due to differences in the shape of the test pieces used, standard shapes and dimensions are adopted. Fig. 11.3 shows the forms of the flat test piece and the round test piece, the following being some of the standard dimensions used for such pieces.

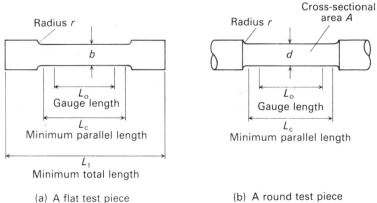

(a) A flat test piece   (b) A round test piece

Fig. 11.3

Flat test pieces

b[mm]	L_o[mm]	L_c[mm]	L_t[mm]	r[mm]
25	100	125	300	25
12·5	50	63	200	25
6	24	30	100	12
3	12	15	50	6

Round test pieces

A[mm²]	d[mm]	L_o[mm]	L_c[mm]	r[mm]	
				Wrought material	Cast material
200	15·96	80	88	15	30
150	13·82	69	76	13	26
100	11·28	56	62	10	20
50	7·98	40	44	8	16

The reason for specifying the above standard forms and dimensions is that if two different size test pieces are taken from the same
piece of material they will, if the test pieces have the above standard
dimensions, give the same stress-strain results. If, however, the dimensions are not those specified as standard then it is likely that
although the test pieces should give the same result they will give

different results because the form and size of a test piece can markedly affect tensile test results.

11.7 Stress-strain graph for mild steel

Fig. 11.4 shows a typical graph for a tensile test on a mild-steel specimen. The stresses have been calculated by dividing the loads by the *original* cross-sectional area of the test piece, and the strains by dividing the extensions by the *original* length selected for the test. The graph differs from a load-extension graph, such as fig. 11.2, only in the matter of scales.

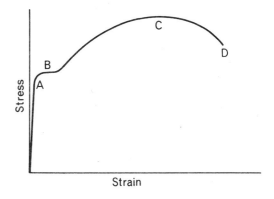

Fig. 11.4 Stress-strain graph for mild steel.

Point A in fig. 11.1 and 11.4 marks approximately the *limit of proportionality* of stress to strain, and is practically the *elastic limit,* i.e. the highest stress that can be applied without producing permanent deformation. For stresses up to the elastic limit, the material returns to its original length on removal of the load.

Portion B of the graph in fig. 11.3 gives the *yield stress,* namely the stress at which, in a tensile test, elongation of the test piece first occurs without increase of load. This plastic yielding is not found in all ductile materials but is a marked feature of the softer irons and steels.

The maximum stress, represented by C in fig. 11.4 is termed the *tensile strength,** and is obtained by dividing the maximum load applied during a tensile test by the *original* cross-sectional area of

* This term replaces *ultimate tensile stress,* the use of which is not now recommended by the British Standards Institution.

the specimen and not by the reduced area of section after plastic extension. At about point C, the cross-sectional area at some point along the length of the test specimen begins to decrease rapidly and the load required to increase the extension beyond C consequently decreases until the specimen ultimately fractures at a load corresponding to point D.

11.8 Percentage elongation

A useful measure of the degree of ductility of a material is the percentage elongation. This involves a measurement of the gauge length of the test piece before the tensile test and after the test is completed and the piece broken, the two broken pieces being put into contact for the measurement.

$$\text{Percentage elongation} = \frac{\text{final length} - \text{initial length}}{\text{initial length}} \times 100 \quad (11.4)$$

Another quantity which is sometimes measured is the percentage reduction in area. The initial cross-sectional area of the test piece is measured and then the smallest cross-sectional area when the test piece has been broken.

$$\text{Percentage reduction in area} = \frac{\text{initial area} - \text{final area}}{\text{initial area}} \times 100$$

Example 11.7 *Taking the elastic limit in fig. 11.1 as* 60 N *and the maximum load as* 154 N, *calculate* (a) *the stress at the elastic limit,* (b) *the tensile strength and* (c) *the percentage final elongation. The wire had an initial length of* 907 mm *and an initial cross-sectional area of* 0·34 mm².

(a) Cross-sectional area = 0·34 mm² = 0·34 × 10⁻⁶ m²,
∴ stress at elastic limit = 60 [N]/(0·34 × 10⁻⁶) [m²]
= 176·5 × 10⁶ Pa = 176·5 MPa.

(b)   Tensile strength = 154 [N]/(0·34 × 10⁻⁶) [m²]
= 453 × 10⁶ Pa = 453 MPa.

(c) In the table given in section 11.4, it is stated that the elongation was 70 mm when the wire broke,

$$\therefore \quad \text{percentage final elongation} = \frac{70 \text{ [mm]}}{907 \text{ [mm]}} \times 100$$
$$= 7 \cdot 7 \text{ per cent.}$$

11.9 Proof stress

The graph of fig. 11.5 is typical of the stress-strain relationship for many materials such as copper, etc. In this diagram, the strain has been plotted as a *percentage* of the original length, i.e. percentage strain $= 100 \times$ elongation/original length.

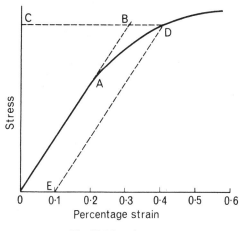

Fig. 11.5 Proof stress.

Point A represents the limit of proportionality for this specimen. Had the strain remained proportional to the stress, the stress-strain relationship would have been represented by the dotted line OB, which is OA extended beyond point A. If we draw a horizontal line CD at a stress OC, then CB represents the percentage strain if the strain had remained proportional to the stress, and CD represents the *actual* percentage strain. Consequently, BD represents the amount by which the percentage strain has departed from proportionality, i.e. BD is the *non-proportional* elongation of the specimen, expressed as a percentage of the original length.

Draw the dotted line DE parallel to OB, then:

$$OE = BD.$$

If OE is, say, 0·1 per cent, then OC is termed the *proof stress* to give a non-proportional elongation equal to 0·1 per cent of the original length of the specimen.

It is usual to express the proof stress as the stresses required to give a non-proportional elongation of 0·1, 0·2 and 0·5 per cent of the original length.

Example 11.8 *A specimen has an initial gauge length of 55 mm and a cross-sectional area of 150 mm². A test on the specimen gave the following results:*

Load [kN]	0	10	20	30	35	38	40
Extension [mm]	0	0·075	0·15	0·23	0·30	0·38	0·6

Determine the 0·2 per cent proof stress in meganewtons per square metre or megapascals.

The graph representing the above table is given in fig. 11.6.

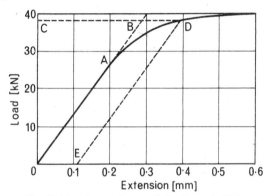

Fig. 11.6 Load-extension graph for Example 10.8

For 0·2 per cent non-proportional extension,

$$\text{actual non-proportional extension} = 55 \text{ [mm]} \times 0·2/100$$
$$= 0·11 \text{ mm.}$$

Hence, in fig. 11.6, from a point E corresponding to an extension of 0·11 mm, draw ED parallel to OA; and from point D, draw a horizontal line to cut the vertical axis at C. Then the *proof load* to give 0·2 per cent non-proportional extension is OC.

From the graph in fig. 11.6, OC = 38 kN = 38 000 N.

Cross-sectional area = 150 mm² = 150 × 10⁻⁶ m²

and   corresponding proof stress $= \dfrac{38\ 000\ [\text{N}]}{150 \times 10^{-6}\ [\text{m}^2]}$

$= 253 \times 10^6\ \text{N/m}^2$

$= 253\ \text{MN/m}^2$ or MPa.

11.10 Tensile strength and factor of safety

The tensile strength of a material has already been referred to as the maximum tensile load (in a test) divided by the original cross-sectional area of the specimen. It is obvious that no such stress must be allowed in any part of a machine or structure. The *permissible stress* should, in fact, always be far below the tensile strength and also almost always below the stress corresponding to the elastic limit. The ratio of the tensile strength to the permissible tensile stress is termed the *factor of safety*.

i.e.   factor of safety $= \dfrac{\text{tensile strength}}{\text{permissible tensile stress}}$ (11.3)

Hence,

permissible tensile stress $= \dfrac{\text{tensile strength}}{\text{factor of safety}}$

The factor of safety to be adopted depends upon many circumstances and it often covers many contingencies very imperfectly known.

The following table gives approximate values of the tensile strength and Young's modulus for several of the most commonly used materials, but materials having a particular name vary widely in their properties and often more precise information is required about the actual materials to be employed. This is particularly true in regard to cast irons and steels of widely differing kinds.

TENSILE STRENGTH AND YOUNG'S MODULUS

Material	Tensile strength in MPa	Young's modulus in GPa
Wrought iron	300 to 400	190
Mild steel	450 to 600	210
Cast iron	120 to 160	120
Copper (hard drawn)	300 to 350	110
Aluminium (hard drawn)	150 to 190	70

Compressive strength of cast iron, 600 to 900 MPa

11.11 Interpreting tensile test data

Fig. 11.7 shows the types of stress strain graphs produced by brittle and ductile materials. A ductile material is one that shows a considerable amount of plastic deformation before it breaks, a brittle material shows virtually no plastic deformation before it breaks. A brittle material will have only a very small percentage elongation, a ductile material will have a much larger value. For example, grey cast iron is a brittle material and has a percentage elongation of about 0·6 per cent. Mild steel is a ductile material and has a percentage elongation of about 30 per cent.

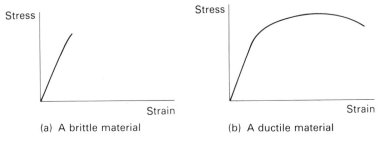

(a) A brittle material   (b) A ductile material

Fig. 11.7 Stress-strain graphs to the breaking point.

Tensile test data are used, not only to determine whether a material is brittle or ductile, but also to test whether materials are to the required specification. The properties of materials can be quite markedly affected by the treatment they receive, either as part of heat treatment or as part of the manufacturing process. A tensile test can thus be used to determine whether the heat treatment of a batch of material has been satisfactorily carried out.

11.12 Compression

Fig. 11.8 shows a short thick bar AB, with opposing forces F pushing or thrusting axially at its ends, i.e. in the opposite directions to the forces which cause tension. In such a case, the material is said to be under *compression* or *compressive stress*. The transmission of forces with action and reaction perpendicular to any cross-section at right angles to the axis is similar to that in the case of tension, but the directions of the forces are reversed. For instance, a short length CD is acted upon by normal *inward* thrusts F across

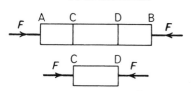

Fig. 11.8 Compressive forces.

the faces C and D as shown in fig. 11.8, and itself exerts *outward* thrusts, each equal to F, on the parts CA and DB.

The average compressive stress across a section is found by dividing force F by the area of cross-section,

i.e. $$\text{compressive stress} = \frac{\text{compressive force}}{\text{cross-sectional area}}$$

The compressive strain is measured as a fraction of the original length,

i.e. $$\text{compressive strain} = \frac{\text{decrease in length}}{\text{original length}}$$

and the modulus of elasticity (direct) or Young's modulus

$$= E = \frac{\text{compressive stress}}{\text{compressive strain}}.$$

The value of Young's modulus for a given material is practically the same for compression as it is for tension.

A tensile force tends to straighten out any kink or curvature in a long rod, but a compressive force tends to increase such imperfections and to buckle the rod. Consequently a structural element used to resist compression, generally called a strut or a column, has to be rather broad in relation to its length, as in the case of tubes and many steel sections such as I, H, L and T shapes.

11.13 Shearing

A material is under the action of *shear stress* across a surface within it when there are forces tending to make the part on one side of the surface slide past the part on the other side of the surface. Thus a pin or rivet shown in fig. 11.9 is subject to shear stress across the plane surface (dotted) AB. The upper part A of the pin tends to slide to the left relative to the lower part B. The tendency of the

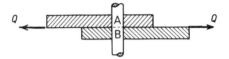

Fig. 11.9 Shear stress.

stress is to *shear* the pin into two parts, A and B. When shears are used for cutting metal or vegetation, actual shearing fracture takes place; but in parts of machines and structures, the *shear stress* is Much below the *shear strength*. The average shear stress in fig. 11.9 is:

$$\frac{\text{shear force}}{\text{area of cross-section between A and B}}$$

A common case of the use of metal to resist shearing is to be found in rivetted joints and in instances where one part of a machine is capable of turning relative to another about a pin. For example, in the knuckle joint shown in fig. 11.10, the pin resists the pull at *two*

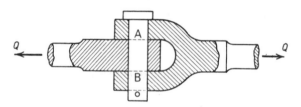

Fig. 11.10 Pin in double shear.

sections, A and B, so that the average shear stress in the pin is equal to force Q divided by *twice* the area of cross-section of the pin. The pin is said to be subjected to *double shear*.

Shear Strain. The form of shear strain which results from a shear stress is illustrated in fig. 11.11. The left-hand view represents a number of piled metal plates which in the right-hand view have been pulled to the right. Suppose these plates were separated one

Fig. 11.11 'Shearing' of plates

from another by layers or films of thick oil. Then the movement to the right due to a pull Q would be resisted by each film of oil and the films would be subject to shear stress. In a continuous material,

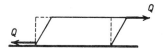

Fig. 11.12 Form of shear strain.

the form will be as illustrated in fig. 11.12, which might represent the visible shear strain produced in rubber. Simple shear strains in metals cannot be easily demonstrated.

11.14 Hardness

The hardness of a material may be defined as the resistance of a material to indentation or scratching. A number of different methods of evaluating hardness have been devised and each of these has developed its own scale of hardness. There is no absolute scale of hardness. With metals the most commonly used tests involve pressing a standard indentor into the surface of the metal and then measurements are made associated with the indentation. The main examples of such tests are the Brinell test, the Vickers test and the Rockwell test.

With the Brinell test a hardened steel ball is pressed, for 10 to 15 s, into the surface of the material being tested by a standard force. After the ball has been removed the diameter of the indentation is measured. The Brinell hardness number is obtained by dividing the size of the force applied by the spherical surface area of the indentation. This area can be obtained by calculation from the measured diameter of the indentation and the diameter of the ball used or by using tables.

Brinell hardness number [HB]
$$= \frac{\text{applied force [kgf]}}{\text{spherical surface area of indentation [mm}^2]}$$

The force unit used is the kgf, this being the weight of a mass of 1 kg and thus being equal to 9·8 N.

The Brinell test cannot be used with very soft or very hard materials. Different ranges of hardness numbers are covered by using different forces and balls of different sizes. The thickness of the material being tested must be at least ten times the depth of the indentation if the results are not to be affected by the thinness of the material.

With the Vickers test a diamond indenter is pressed into the surface for 10 to 15 s by a standard force. This indenter gives rise to a pyramid-shaped indentation. After the indenter has been removed the diagonals of the indentation are measured. The Vickers hardness number is obtained by dividing the force by the surface area of the indentation. The surface area can be calculated, or derived using tables, from the shape of the indenter and the lengths of the diagonals.

Vickers hardness number [HV]

$$= \frac{\text{load [kgf]}}{\text{sloping area of the indentation [mm}^2\text{]}}$$

As the diamond used always gives an indentation in the form of a right pyramid with a square base and a vertex angle of 136°, the sloping area is always $d^2/1{\cdot}854 \text{ mm}^2$, where d is the average of the two diagonal measurements.

The Vickers test has an advantage over the Brinell test in that it is easier to determine the lengths of the diagonals than the diameter of a circle and so the test is more accurate. Otherwise it is similar to the Brinell test.

The Rockwell test uses either a diamond cone of a hardened steel ball as indentor, both giving rise to spherical indentations. A force is used to press the indentor into the surface. When equilibrium has been reached an indicator which responds to the depth of penetration is set to its zero position. Without removing this force a further force is applied with the resulting increase in the depth of penetration. When equilibrium is reached this additional force is removed. This allows a partial recovery of the material and the depth of penetration is reduced. This is due to the deformation of the material not being entirely plastic. The indicator is then read for this final depth of penetration.

Rockwell hardness number [HR] $= E - e$

where E is a constant determined by the form of the indenter and e is this extra permanent penetration.

A variety of indenters and forces are used to give several hardness scales, the scale used being denoted by a letter. The B and C scales are probably the most commonly used. The B scale involves a $1{\cdot}588 \text{ mm}$ diameter steel ball being used with an additional force of $0{\cdot}98 \text{ kN}$ and is used with copper alloys, aluminium alloys and soft steels. The C scale involves the use of a diamond with an additional force of 1·47 kN and is used for steels, hard cast irons and deep case-hardened steels.

The Rockwell test is more suited to 'on site' or general workshop hardness measurements than the Brinell or the Vickers test because it does not involve having flat or polished surfaces, as they do, because of the length measurements that have to be made. It gives only a very small indentation, smaller than the Brinell or Vickers tests, and so is more likely to be affected by local hardness variations within a material. As with the other tests, it is not suitable for hardness measurements on thin sheet. However a special version of the test exists which can be used with such thicknesses.

There is an approximate relationship between hardness values and tensile strength. For a particular hardness scale

$$\text{tensile strength} = \text{a constant} \times \text{hardness value}$$

For the Brinell hardness scale the constant is 3·54 for annealed steels, 3·24 for quenched and tempered steels, 5·6 for brass and 4·2 for aluminium alloys.

Summary of Chapter 11

Stress is the force per unit area of a solid body.
Elastic strains are those which disappear on the removal of stress. For most materials worked within the elastic limit, the strain is proportional to the stress.

$$\text{Tensile (or compressive) stress} = \frac{\text{tensile (or compressive) force}}{\text{cross-sectional area}} \quad (11.1)$$

$$\text{Tensile (or compressive) strain} = \frac{\text{change in length}}{\text{original length}} \quad (11.2)$$

Modulus of elasticity (direct) or Young's modulus

$$= E = \frac{\text{tensile (or compressive) stress}}{\text{tensile (or compressive) strain}} \quad (11.3)$$

$$\text{Percentage elongation} = \frac{\text{final length} - \text{initial length}}{\text{initial length}} \times 100 \quad (11.4)$$

Proof stress is the tensile stress that produces a non-proportional elongation equal to a specified percentage of the original length.

$$\text{Tensile strength} = \frac{\text{maximum tensile force}}{\text{original cross-sectional area}}$$

$$\text{Factor of safety} = \frac{\text{tensile strength}}{\text{permissible tensile stress}} \qquad (11.5)$$

$$\text{Shear stress} = \frac{\text{shear force}}{\text{cross-sectional area}}$$

$$\text{Shear strength} = \frac{\text{maximum shear force}}{\text{cross sectional area}}$$

EXAMPLES 11

1. What is the stress, in megapascals, in a wire, 1 mm diameter, under a load of 120 N?
2. Calculate the diameter of a steel rod (to the nearest millimetre) to carry a tensile load of 90 kN, if the stress is not to exceed 100 MPa.
3. What pull, in kilonewtons, can be borne by a tie rod, 40 mm diameter, if the stress is limited to 120 MPa?
4. A cylindrical concrete vertical pillar, having a diameter of 250 mm, supports a load of 10 kN. Calculate the stress in megapascals.
5. Convert:
   (a) 20 N/mm^2 into megapascals;
   (b) 500 kPa into millinewtons per square millimetre;
   (c) 80 kN/mm^2 into gigapascals;
   (d) 50 GPa into kilonewtons per square millimetre;
   (e) 400 Pa into micronewtons per square millimetre.
6. If a spring is elastically stretched 60 mm by a load of 72 n, how much will it be stretched by a load of 40 N?
7. A helical spring carrying load of 600 N is compressed by 20 mm. What would be the load required to compress the spring by 8 mm?
8. A steel rod, 10 mm diameter and 3 m long, stretches 1·8 mm for a pull of 10 kN. Calculate the value of Young's modulus for the steel.
9. A steel rod, 20 mm diameter, is subjected to a pull of 40 kN. Calculate (a) the stress, (b) the strain and (c) the total extension on a length of 200 mm. Assume $E = 200$ GPa.
10. A body having a mass of 1 Mg (or 1 t) is attached to the lower end of a vertical aluminium rod, 30 mm diameter and 2 m long. Calculate (a) the stress in the rod and (b) the extension, assuming $E = 70$ GPa.
11. When a metal block is suspended by a steel wire, 5 m long and 2·5 mm diameter, the elongation is 3 mm. Calculate the mass of the block, assuming $E = 200$ GPa.

12. Calculate the diameter of a steel rod to carry a load of 8 kN if the extension is not to exceed 0·04 per cent. *Assume E* = 210 GPA.

13. The following results were obtained on a mild-steel specimen having an initial gauge length of 50 mm and an initial cross-sectional area of 160 mm² :

Load [N]	50	75	100	125	150
Extension [mm]	0·48	0·73	0·97	1·20	1·44

Draw a load-extension graph and calculate the modulus of elasticity (direct) of the material.

14. In an experiment on a 2-m length of 1-mm diameter steel wire, the following results were obtained:

Load [N]	50	75	100	125	150
Extension [mm]	0·48	0·73	0·97	1·20	1·44

Calculate Young's modulus for the steel.

If a tie-bar in a bridge is made of the same material and is 4 m long, what will be the extension when the stress is 100 MPa?

15. The following results were obtained on a rod having a diameter of 12·5 mm and a length of 250 mm:

Load [kN]	12·5	25	37·5	40	41	42·5	45
Extension [mm]	0·12	0·24	0·36	0·55	1·8	3·85	5·1

Load [kN]	47·5	50	52·5
Extension [mm]	6·3	7·8	9·5

Plot load-extension graphs, using a scale of 1 mm to 0·5 kN and 1 mm to 0·02 mm for extensions up to 2 mm and 1 mm to 0·1 mm for extensions up to 10 mm. From these graphs, determine (*a*) the approximate value of the elastic limit, (*b*) Young's modulus and (*c*) the 0·2 per cent proof stress.

16. The following results were obtained from a tensile test on a copper wire having a diameter of 1·3 mm and a length of 2·3 m:

Load [N]	25	50	75	100	125	150	160	170
Extension [mm]	0·38	0·75	1·15	1·53	2·8	8·6	15·0	28·2

Plot the load-extension graph and determine (*a*) Young's modulus and (*b*) the 0·5 per cent proof stress.

17. A steel tie-rod, 10 m long and 3 mm diameter, is subjected to a pull of 600 N. Calculate (*a*) the stress, (*b*) the elongation if the modulus of elasticity (direct) is 210 GPa and (*c*) the factor of safety if the tensile strength is 500 MPa.

18. The overhead conductor of an electric transmission line is a hard-drawn copper wire, 10 mm diameter, having a tensile strength of 320 MPa and a Young's modulus of 110 GPa. Assuming a factor of safety of 6, calculate (*a*) the maximum allowable pull on the wire and (*b*) the corresponding elongation on a span of 40 m.

19. A wrought-iron vertical pillar, 1·3 m long, has a cross-sectional area of 2000 mm². The pillar supports a load of 80 kN. Assuming *E* = 180 GPa, calculate the compression.

20. A hollow cast-iron vertical cylinder, 3 m long when unloaded, has an outer diameter of 150 mm and an inner diameter of 130 mm. Calculate (a)· the maximum load that can be supported by the cylinder if the stress is not to exceed 80 MPa and (b) the corresponding decrease in length. Assume $E = 100$ GPa.
21. Two lengths of flat tie-bar (fig. 10·7) are connected together by a lap-riveted joint with three rivets, each 10 mm diameter. Calculate the maximum pull that can be applied to the tie-bars if the shear stress in the rivets is not to exceed 60 MPa.
22. If a knuckle-joint (fig. 10.8) is subjected to a pull of 100 kN, calculate the diameter of pin necessary in order that the average shear stress shall not exceed 80 MPa.
23. A mass of 5 kg is suspended at the end of a wire having a diameter of 2 mm. Calculate the stress in the wire, in newtons per square millimetre.
24. Calculate the mass, in tonnes, that is supported by a cylindrical vertical pillar, 1·5 m in diameter, when the stress is 20 N/mm².

ANSWER TO EXAMPLES 11

1. 153 MPa.
2. 34 mm.
3. 150·8 kN.
4. 0·204 MPa.
5. 20 MPa; 500 mN/mm²; 80 GPa; 50 kN/mm²; 400 μN/mm².
6. 33·3 mm.
7. 240 N.
8. 212 GPa.
9. 127·5 MPa, 6·375 × 10⁻⁴, 0·1275 mm.
10. 13·9 MPa, 0·397 mm.
11. 60 kg.
12. 11 mm.
13. 234 GPa.
14. 265·5 GPa, 1·506 mm.
15. About 306 MPa, 212 GPa, about 330 MPa.
16. 114 GPa, about 119 MPa.
17. 84·9 MPa, 4·04 mm, 5·9.
18. 4·19 kN, 19·4 mm.
19. 0·289 mm.
20. 352 kN, 2·4 mm.
21. 14·1 kN.
22. 28·2 mm.
23. 15·6 N/mm².
24. 3600 tonnes.

CHAPTER 12

Fluid Pressure

12.1 Fluids

Matter may be divided into three states: (a) solid; (b) liquid; (c) gaseous. Matter in the liquid and gaseous states can flow and is therefore referred to as a *fluid*.

The *pressure* exerted by a fluid on the walls of the containing vessel is the force per unit area exerted by the fluid, and the SI unit of pressure is the *newton per square metre* or *pascal*.

12.2 Pressure

When the tyre of a bicycle is inflated, the compressed air exerts an outward thrust on the inner surface of the tyre, and the thrust per unit area is termed the *pressure*. The SI unit of pressure is the *newton per square metre* or *pascal*, exactly the same as for stress.

The atmospheric pressure known as the *standard atmospheric pressure* is 101 325 Pa or N/m^2.

The term *pressure* generally refers to the thrust per unit area exerted by *fluids* such as water, air, steam, etc., whereas the term *stress* refers to the force per unit area exerted in a solid body. Sometimes the term 'pressure' is used to denote the total thrust on a surface; but to avoid confusion, it is better in such a case to use the term 'thrust' or 'force'.

Example 12.1 *Steam at a pressure of* 1·5 *MPa is applied to a piston having a diameter of* 200 *mm. Calculate* (a) *the thrust on the piston and* (b) *the work done when the piston moves through a distance of* 350 *mm, assuming the steam pressure to remain constant.*

(a) $\qquad$ Area of piston $= (\pi/4) \times (200)^2 = 31\ 400$ mm^2
$$= 0{\cdot}0314 \text{ m}^2.$$

$\qquad$ Pressure on piston $= 1{\cdot}5$ MPa $= 1{\cdot}5 \times 10^6$ Pa,

∴ $\qquad$ thrust on piston $= 1{\cdot}5 \times 10^6$ [Pa] $\times 0{\cdot}0314$ [m^2]
$$= 47\ 100 \text{ N} = 47{\cdot}1 \text{ kN}.$$

(b)   Displacement of piston = 350 mm = 0·35 m,

∴   work done = 47 100 [N] × 0·35 [m]

                         = 16 500 J = 16·5 kJ.

12.3 Variation of pressure with depth in a liquid

Suppose the tank in fig. 12.1 to be nearly filled with a liquid having
a density of ρ kilograms per cubic metre. Let us consider a vertical

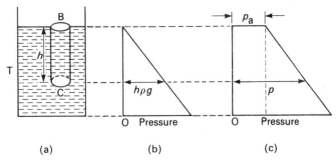

Fig. 12.1 Variation of pressure with depth in a liquid.

cylindrical column BC of height h metres and cross-sectional area a
square metres. If the upper surface B is on a level with the surface
of the liquid, and if the effect of atmospheric pressure on the surface
of B is neglected, the only vertical force acting downward on the
bottom surface at C is the *weight* of the liquid enclosed in cylinder
BC.

Volume of liquid in cylinder BC = ah cubic metres,

∴   mass of liquid in cylinder BC = $ah\rho$ kilograms.

If g is the gravitational force, in newtons, on a mass of 1 kg,

weight of liquid in cylinder BC = $ah\rho g$ newtons

and pressure on lower surface of BC = $ah\rho g/a = h\rho g$ pascals (12.1)

Since the liquid is assumed stationary, the downward pressure at
C is balanced by an equal upward pressure. This equilibrium applies
to every pair of forces acting in opposite directions throughout the
liquid. Hence, if the atmospheric pressure on the surface of the liquid
is neglected, *the pressure is directly proportional to the depth*, as
shown in fig. 12.1(b).

If the atmospheric pressure on the surface of the liquid is p_a, then the total pressure p at depth h below the surface is given by:

$$p = p_a + h\rho g \tag{12.2}$$

as shown in fig. 12.1(c).

12.3 Properties of fluid pressure

(a) *The pressure exerted at any point in a stationary fluid is the same in all directions*, as shown by the arrows in fig. 12.2. If this

Fig. 12.2 Pressure on a point in a fluid.

were not the case, the forces acting on an element of the fluid, represented by the dot, would not be in equilibrium, and the resultant force available would cause the element to move.

(b) *The pressure exerted in a stationary fluid is the same at all points in the same horizontal plane.* Suppose a vessel to be constructed as shown in fig. 12.3 and to be nearly filled with a liquid. Then, if the liquid is stationary, the surface of the liquid is at the same level in each of the containers P, Q, R and S. Similarly, the pressure at a *given vertical* distance below the surface is the same for the four

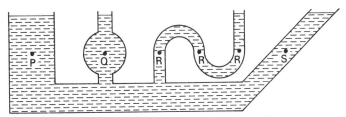

Fig. 12.3 Pressure at points on same horizontal level.

containers. A simple example is the case of a teapot where the level of tea in the pot is exactly the same as that in the spout.

It is this property of a liquid that enables water to be transmitted from a reservoir constructed on an elevation that is higher than that of the district or town to be supplied. The level of the pipes may rise or fall according to the contour of the ground, but no point of the pipes must rise above the level of the surface of the water in the reservoir, otherwise no water will pass that point.

(*c*) *The pressure exerted by a stationary fluid on a solid surface is normal (or perpendicular) to that surface.* If this were not the case, the fluid would move relative to the solid surface.

Example 12.2 *A rectangular tank, 5 m × 2 m, is 4 m high. It is filled with oil having a relative density of 0·8. Calculate (a) the pressure at a depth of 2 m, (b) the pressure and the total thrust on the bottom of the tank and (c) the total thrust on one of the end surfaces. Neglect the effect of atmospheric pressure.*

(*a*) Since density of water = 1000 kg/m³

∴   density of oil = 0·8 × 1000 = 800 kg/m³.

From expression (12.1)

pressure at depth of 2 m = 2 [m] × 800 [kg/m³] × 9·81 [N/kg]
$$= 15\,696 \text{ Pa or N/m}^2$$
$$\simeq 15\cdot7 \text{ kPa or kN/m}^2.$$

(*b*) Pressure on bottom of tank = 4 [m] × 800 [kg/m³] × 9·81 [N/kg]
$$= 31\,392 \text{ Pa or N/m}^2$$
$$\simeq 31\cdot4 \text{ kPa or kN/m}^2.$$

Area of bottom of tank = 5 [m] × 2 [m] = 10 m²

∴ total thrust on bottom of tank $\Big\}$ = 31·4 [kN/m²] × 10 [m²]
$$= 314 \text{ kN}.$$

(*c*) Area of one end surface   = 2 [m] × 4 [m] = 8 m²

From fig. 12.1(b), it follows that the average pressure on each of the vertical surfaces occurs at a depth of 2 m and has already been found to be 15·7 kPa.

∴ total thrust on one end surface = 15·7 [kN/m²] × 8 [m²]
$$= 125\cdot6 \text{ kN}.$$

Example 12.3 *The tank, shown in elevation in fig. 12.4, is 3 m broad (in a direction perpendicular to the diagram) and has a vertical height of 4 m. End surface AB has a slope of 60° to the horizontal. The tank is open to the atmosphere and is filled with water having a density of*

1000 kg/m³. The atmospheric pressure is 101·3 kN/m². Calculate the
total thrust on surface AB.

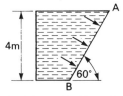

Fig. 12.4 Diagram for Example A.3.

The average pressure on surface AB occurs at a *vertical* depth of
2 m. Hence, from expression (12.1),

$$\left. \begin{array}{l} \text{average pressure due} \\ \text{to water only} \end{array} \right\} = 2 \text{ [m]} \times 1000 \text{ [kg/m}^3] \times 9 \cdot 81 \text{ [N/kg]}$$

$$= 19\,620 \text{ N/m}^2 = 19 \cdot 62 \text{ kPa or kN/m}^2.$$

Hence the actual pressure increases uniformly from 101·3 kPa at
the surface of the water to (101·3 + 39·24), namely 140·54 kPa at
the bottom of the tank. This pressure is normal to the sloping
surface, as indicated by the arrows in fig. 12.4.

$$\therefore \left. \begin{array}{l} \text{average pressure normal} \\ \text{to surface AB} \end{array} \right\} = (101 \cdot 3 + 140 \cdot 54)/2$$

$$= 120 \cdot 92 \text{ kPa.}$$

Length AB of sloping surface $= 4/\sin 60° = 4/0 \cdot 866$

$$= 4 \cdot 62 \text{ m}$$

$\therefore$ area of sloping surface $= 4 \cdot 62 \text{ [m]} \times 3 \text{ [m]} = 13 \cdot 86 \text{ m}^2$

$$\left. \begin{array}{l} \text{and total thrust on sloping} \\ \text{surface} \end{array} \right\} = 120 \cdot 92 \text{ [kN/m}^2] \times 13 \cdot 86 \text{ [m}^2]$$

$$= 1676 \text{ kN} = 1 \cdot 676 \text{ MN.}$$

12.5 Measurement of gauge pressure

Gauge pressure is the difference between the *absolute pressure* and the
atmospheric pressure:

i.e. absolute pressure = gauge pressure + atmospheric pressure.

The simplest method of measuring the pressure of a gas in a

container C is to attach to the container one limb A of a U-tube
containing a liquid, as shown in fig. 12.5. The other limb B is open

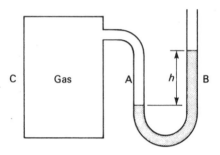

Fig. 12.5 manometer.

to the atmosphere. This type of pressure gauge is referred to as a
manometer.

When the gas pressure in C is greater than the atmospheric
pressure, the level of the liquid is forced down in limb A and up in
limb B. If the difference in the levels is h and the density of the
liquid is ρ, it follows from expression (12.1) that the difference
between the pressure p exerted by the gas in C and the atmospheric
pressure p_a is equal to the pressure exerted by a column of height h
of the liquid, namely $h\rho g$,

$$\text{i.e. gauge pressure} = p - p_a = h\rho g \qquad (12.3)$$

If p were equal to, say, $0.8\,p_a$, and if p_a were $101\,325\,\text{Pa}^2$,
namely the *standard* value of the atmospheric pressure, then substituting in expression (12.3), we have:

$$9.81 \times h\rho = (0.8 \times 101\,325) - 101\,325$$
$$= -20\,265\,\text{Pa or N/m}^2.$$

If the liquid in the manometer is water, $\rho = 1000\,\text{kg/m}^3$,

$$\therefore \qquad 9.81\,[\text{N/kg}] \times h \times 1000\,[\text{kg/m}^3] = -20\,265\,[\text{N/m}^2]$$
$$\text{and } h = -2.07\,\text{m}.$$

The minus sign indicates that the level of the water in limb A
is *above* that in limb B by 2·07 m and that the pressure in vessel C
is less than the atmospheric pressure. Such a U-tube would be
inconveniently large. If mercury (relative density = 13·6) were

substituted for water, ρ would be 13 600 kg/m^3 and h would be $-2\cdot07/13\cdot6$, namely $-0\cdot152$ m. This means that the level of the mercury in limb A would be 15·2 cm above that in limb B and the size of the manometer would be much more convenient.

12.6 The Bourdon gauge

The Bourdon gauge is a very commonly used instrument for the measurement of gauge pressures. The gauge consists of a metal tube of approximately elliptical cross-section, the tube being in the form of a C (fig. 12.6). One end of the tube is fixed and connected to the pressure system being measured. The other end of the tube is sealed and free to move, the movement being communicated to a pointer via a lever and a rack and pinion. When the pressure in the tube increases the tube tends to straighten out. This movement is magnified and transmitted to the pointer by the lever and rack-and-pinion arrangement.

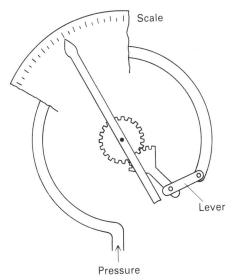

Fig. 12.6 The principles of the Bourdon gauge.

Gauge pressure, i.e. pressure differences between that in the tube and outside the tube, can be measured from about 10^3 Pa to 10^8 Pa, depending on the form of the tube used. It is a relatively cheap instrument and robust.

12.7 Measurement of atmospheric pressure

A glass tube T, about 80 cm long and sealed at one end, is *filled*
with mercury and the open end is then dipped into mercury contained in a vessel V, as in fig. 12.7. The height of the mercury column
falls to a value h, leaving a vacuum A. Such an arrangement is termed
a *barometer*.

The column of mercury in T is maintained by the atmospheric
pressure p_a on the surface of the mercury in V. In other words, for
a point in the column of mercury on a level with the surface of the
mercury in vessel V, the pressure is equal to the atmospheric pressure.
Assuming $p_a = 101\,325\,\text{Pa}^2$ and $\rho = 13\,600\,\text{kg/m}^3$, then from expression (12.1):

$$9\cdot81\,[\text{N/kg}] \times h \times 13\,600\,[\text{kg/m}^3] = 101\,325\,[\text{N/m}^2]$$
$$\therefore \qquad\qquad h = 0\cdot76\text{ m} = 76\text{ cm}.$$

i.e. a vertical column of mercury, 76 cm high, exerts a pressure equal
to the standard value of the atmospheric pressure.

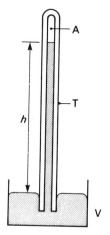

Fig. 12.7 A barometer.

Summary of chapter 12

Pressure is force per unit area. The SI unit is the pascal (Pa) or Nm^{-2}.

$$\text{Pressure} = h\rho g \qquad\qquad (12.1)$$

Absolute pressure = gauge pressure + atmospheric pressure.

EXAMPLES 12

1. A rectangular tank is 20 m long and 5 m wide. It contains water to a depth of 6 m. Calculate (*a*) the pressure, in kilopascals, at a depth of 3 m, and (*b*) the total thrust, in meganewtons, on (i) the bottom of the tank, (ii) one of the long sides of the tank. Neglect the effect of atmospheric pressure.

2. Calculate the absolute pressure, in pascals, at a depth of 300 m in sea water, assuming the relative density of sea water to be 1·03. If a square plate, 3 m wide, is immersed horizontally at a depth of 300 m, what is the downwards thrust, in meganewtons, on the top side of the plate? Assume the atmospheric pressure to be 101 kPa.

3. A lock gate has a depth of 4 m of fresh water on one side and is dry on the other side. It is 8 m wide. What is the net thrust on the gate?

4. A rectangular plate, 3 m × 2 m, is immersed in oil, with its plane at an angle of 30° to the vertical. Its upper 3-m edge is horizontal and 1 m below the surface of the oil. The relative density of the oil is 0·7. Calculate the average absolute pressure on the plate, in kilopascals, and the total thrust on one side of the plate, in kilonewtons. Assume the atmospheric pressure to be 100 kPa.

5. A layer of oil, 1 m thick, rests on water. The relative density of the oil is 0·8. Calculate the gauge pressure, in kilopascals, at a depth of 2 m below the lower surface of the oil.

6. A U-tube pressure gauge contains mercury. The level of the mercury in the limb open to the atmosphere is 10 cm higher than that in the other limb. The relative density of mercury is 13·6. Calculate the value of the gauge pressure, in pascals. If the atmospheric pressure is 101 kPa, what is the absolute pressure of the gas?

7. If the height of the mercury column in a certain barometer is 720 mm, what is the corresponding value of the atmospheric pressure, in pascals? Assume the relative density of mercury to be 13·6.

ANSWERS TO EXAMPLES 12

1. 29·43 kPa; 2·943 MN, 1·766 MN.
2. 3·13 MPa; 28·2 MN.
3. 628 kN.
4. 112·8 kPa; 677 kN.
5. 27·47 kPa.
6. 13·3 kPa; 1143 kPa.
7. 96·1 kPa.

CHAPTER 13

Heat and temperature

13.1 Heat

References have been made in previous chapters to the fact that work done in overcoming friction is converted into heat. There are numerous everyday actions that involve the same principle, e.g. rubbing one's hands together, striking a match, applying the brake on a vehicle, etc. In each case, the mechanical energy expended reappears in the form of thermal energy and produces an increase of temperature.

In order to appreciate what is meant by thermal energy, we have to consider the molecular structure of matter. All substances consist of molecules, the molecule of an element (e.g. hydrogen or copper) or of a compound (e.g. water or common salt) being the smallest particle of the element or compound that can normally exist alone. In a solid, the molecules are arranged in an orderly manner. Their mean positions relative to one another are fixed, but they oscillate about their mean positions at a speed that increases with increase in temperature. Hence molecules of solids possess kinetic energy owing to their vibration about their mean positions.

In a liquid, the molecules are able to move about at random. In a gas, the number of molecules in a given space is far less than in a liquid; consequently they are much freer to move about at random than they are in a liquid. The higher the temperature of the liquid or the gas, the greater is the speed of movement of the molecules. It follows that thermal energy is stored in a substance in the form of kinetic energy of the molecules of which the substance is composed. The gain of heat by a material is usually accompanied by the following effects:

(a) an increase in the temperature of the material,
(b) an increase in the volume of the material.

There may also be a change in the state of the material; for instance, a solid such as ice may be melted, or a liquid such as water may be vaporized.

13.2 Heat transfer

Heat is the transfer of energy from a hot body to a cooler body in one or more of the following ways:

(*a*) by conduction,
(*b*) by convection,
(*c*) by radiation.

(*a*) *Conduction.* Conduction of heat takes place between bodies in actual contact if they are at different temperatures. It also takes place between parts of the same body if there is a difference of temperature between these parts.

The thermal conductivity of different materials varies widely; for instance, if one end of a glass rod, say 10 cm long, was heated to a high temperature, the other end could be held in the hand without any discomfort; but this would not be possible with a metal rod of the same length. In other words, metals are good conductors of heat, whereas materials such as glass, cork, cotton wool and gases are very poor conductors.

(*b*) *Convection.* Convection is the conveyance of heat by the actual movement of a hot fluid which may be a liquid or a gas. In most cases, convection takes place automatically; for example, in a hot-water heating system, water heated by a boiler expands so that its density becomes lower than that of the cold water. Consequently the hot water rises and thereby produces circulation of water in the pipes, enabling cold water to move towards the source of heat. Thus heat is transmitted by convection from the boiler to the place where that heat can be utilized, e.g. for warming a room.

Similarly, when air is heated by, say, a lamp or a radiator, it expands, and its consequent lower density causes it to rise, thereby producing an upward draught in the vicinity of the source of heat and a downward movement of air elsewhere in the room.

(*c*) *Radiation.* Radiation is the transmission of heat by wave or vibratory motion in the space between the source and the body on which the waves impinge. The nature of heat waves is the same as that of light waves and of radio waves—the only difference is the frequency of the waves, i.e. the number of vibrations or oscillations per second. Thermal radiation can be transmitted through a vacuum; i.e. the transference of energy by radiation is not dependent upon a material medium, as is the case with conduction and convection. For example, thermal energy from the sun reaches the earth's atmosphere by radiation through empty space.

Thermal radiation, when not impeded by any object, travels in straight lines, obeys the same laws as light and is therefore reflected from a polished metal surface. For a given surface temperature, the heat radiated per unit area is much less if that surface is polished than if it is black. As the temperature of a body is raised, the frequency range of the radiated energy increases, and a point is reached at which the body becomes luminous, i.e. it radiates visible (or light) energy as well as thermal energy.

13.3 Temperature

The temperature of a body relates to its degree of hotness or coldness and is independent of the size and physical nature of the body. Two bodies are said to be at the same temperature if no heat flows from one body to the other when they are brought into contact with each other.

The sense of touch is a very unreliable guide as to the temperature of a body. For instance, if we were to touch two plates—one of iron and the other of wood—which had been immersed for a few minutes in hot water, the iron plate would *feel* hotter than the wooden plate, though they were actually at the same temperature. This is due to the fact that heat is conducted far better by iron than by wood. It is therefore necessary to use an instrument, called a *thermometer*, to measure the temperature of a body.

Of the various types of thermometer, the mercury thermometer is by far the most commonly used for measuring temperatures ranging between the freezing and boiling points of water. The mercury thermometer consists of a glass rod G (fig. 13.1), having a small uniform bore C. At one end of the rod is a bulb B having a very thin glass wall so that heat can pass easily to and from the mercury contained in the bulb. For a given rise of temperature, mercury expands more than glass; consequently, when the mercury thermometer is placed in contact with a hot body, the expansion of the mercury in bulb B causes the level to rise in bore C, and a scale alongside C can be calibrated to read the temperature.

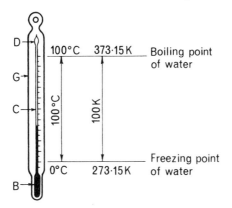

Fig. 13.1 Mercury thermometer.

The upper end of G is sealed when the mercury is heated sufficiently to fill the bore, so that when the mercury cools, the space above the mercury is a vacuum. A small bulb D is provided to prevent breakage of the thermometer due to the temperature rising a little above that for which the thermometer is calibrated.

13.4 Temperature scales

The SI unit of temperature is the *kelvin** (K) and the corresponding scale of temperature is referred to as the *thermodynamic temperature scale*. This scale is based upon thermodynamic considerations that are well beyond the scope of this book. All we need to state here is that, on this scale, the temperature at which ice melts under standard atmospheric pressure is exactly 273·15 K, as shown in fig. 13.1.

The thermodynamic scale suggests that if the temperature of a body could be reduced to zero, the body would possess no thermal energy. Such a condition can never be attained in practice.

A temperature change of one kelvin is precisely the same as the temperature change of one degree Celsius which is the unit of the

* The name 'kelvin' commemorates Lord Kelvin (1824–1907), a famous British scientist and inventor who introduced the dynamical theory of heat.

Celsius (or centigrade) scale suggested by Anders Celsius, a Swedish physicist and astronomer, in 1742. On this scale, the temperature at which ice melts is taken as zero (0°C) and that at which water boils under standard atmospheric pressure is taken as 100°C, as shown in fig. 13.1. In other words, one degree Celsius is one-hundredth of the difference between the temperature at which ice melts (or water freezes) and that at which water boils under standard atmospheric pressure.

The temperature of melting ice under standard atmospheric pressure can therefore be stated as 0°C or 273·15 K and that of boiling water under standard atmospheric pressure can be stated as 100°C or 373·15 K. In general, if t be the temperature of a body in degrees Celsius and T be the same temperature in kelvins, then:

$$t = T - 273·15 \qquad (13.1)$$

It should be noted that °C or K, can be used to denote temperature interval and also the number of temperature units above or below a specified datum. For example, a temperature rise from, say, 20 degrees Celsius to 50 degrees Celsius can be expressed as 30°C or 30 K. If the temperature of a material is, say, 10 degrees Celsius, this can be expressed as 10°C or 283·15 K.

13.5 Quantity of heat

Since heat is a form of energy, the SI unit of heat is the *joule*, a quantity already discussed in sections 1.10 and 9.1. In practice, however, it is often more convnient to use the *kilojoule* (kJ) or the *megajoule* (MJ) as the unit of thermal energy.

Unfortunately, the amount of heat required to raise the temperature of a given mass of water by 1°C varies slightly over the 0–100°C range. The following table gives the heat required to raise the temperature of 1 kg of water by 20°C over this range:

Temperature range	Heat required
0–20°C	83 900 J/kg
20–40°C	83 600 J/kg
40–60°C	83 600 J/kg
60–80°C	83 800 J/kg
80–100°C	84 200 J/kg

It follows from the above data that, for most purposes, we can assume the heat required to raise the temperature of 1 kg of water by 1°C to be 4190 J or 4·19 kJ.

13.6 Specific heat capacity

Different substances absorb different amounts of heat to raise a given mass of the substance by one degree. The quantity of heat required to raise the temperature of 1 kg of the substance by 1°C is termed the *specific heat capacity* of that substance. For instance, it was stated in section 13.5 that the quantity of heat required to raise the temperature of 1 kg of water by 1°C is approximately 4190 J. Hence we can say that the specific heat capacity of water is 4190 J/kg °C. It follows that if the temperature of m kilograms of water is raised by t degrees,

$$\text{quantity of heat required} = 4190\,mt \text{ joules} \qquad (13.2)$$

In general, if the specific heat capacity of a substance is c joules per kilogram degree Celsius, the heat required to raise the temperature of m kilograms of the substance by t degrees

$$= mct \text{ joules} \qquad (13.3)$$

The following table gives approximate values of the specific heat capacity for some well-known substances for a range of temperature between 0°C and 100°C:

Substance	Specific heat capacity
Water	4190 J/kg °C
Ice (0°C to −20°C)	2100 J/kg °C
Copper	390 J/kg °C
Iron	500 J/kg °C
Aluminium	950 J/kg °C
Brass	370 J/kg °C
Dry air at standard atmospheric pressure	1015 J/kg °C

Example 13.1 *Calculate the quantity of heat required to raise the temperature of 6 kg of water from 10°C to 25°C.*

Increase of temperature = 25 − 10 = 15°C.

Substituting in expression (13.2), we have:

$$\text{heat required} = 4190 \text{ [J/kg °C]} \times 6 \text{ [kg]} \times 15 \text{ [°C]}$$
$$= 377\,000 \text{ J} = 377 \text{ kJ}.$$

Example 13.2 *How many kilograms of copper can be raised from 15°C to 60°C by the absorption of 80 kJ of heat? Assume the specific heat capacity of copper to be 390 J/kg °C.*

Increase of temperature $= 60 - 15 = 45°C$.

Quantity of heat $= 80 \text{ kJ} = 80\,000 \text{ J}$.

Substituting in expression (13.3), we have:

$$80\,000 \text{ [J]} = m \times 390 \text{ [J/kg °C]} \times 45 \text{ [°C]}$$
$$\therefore \quad m = 4\cdot56 \text{ kg}.$$

Example 13.3 *Calculate the mass of aluminium that can be raised in temperature from 10°C to 150°C by the heat released from burning 1 m³ of fuel gas. Assume that the heat obtainable from 1 m³ of the gas is 17 MJ, and that 75 per cent of this heat can be made effective. The specific heat capacity of aluminium is 950 J/kg °C.*

Increase of temperature $= 150 - 10 = 140°C$.

$$\left.\text{Effective heat per cubic} \atop \text{metre of gas}\right\} = 17 \times 0\cdot75 = 12\cdot75 \text{ MJ}$$
$$= 12\cdot75 \times 10^6 \text{ J}.$$

Let m be the mass of the aluminium, in kilograms.

Substituting in expression (13.3), we have:

$$12\cdot75 \times 10^6 \text{ [J]} = m \times 950 \text{ [J/kg °C]} \times 140 \text{ [°C]}$$
$$\therefore \quad m = 95\cdot9 \text{ kg}.$$

13.7 Heat transfer in mixtures

When a hot substance is mixed or brought into contact with a cooler one, heat is transferred from the hotter to the cooler substance until the two are ultimately at the same temperature, which is below the original temperature of the hotter substance and above that of the cooler one. One of the substances is frequently a liquid; sometimes both are. It is then necessary to use some sort of containing vessel, and this vessel as well as the liquid will be heated or cooled. The heat thus gained or lost by the vessel should be taken into account.

Calculations on mixtures of hot and cold substances are based on the principle that the heat lost by the hot substance is equal to that gained by the cold substance; i.e. it is assumed that no loss of heat occurs to any external substance such as the surrounding air.

Example 13.4 *A metal vessel and its water contents are together equivalent to 6 kg of water and their temperature is 8°C. If 5 kg of water at 35°C are added, what will be the resulting temperature of the mixture?*

If t be the resulting temperature, then from expression (13.2),
total *gain* of heat by the cold water and vessel
$$= 4190 \text{ [J/kg °C]} \times 6 \text{ [kg]} \times (t - 8) \text{ [°C]}$$

and total *loss* of heat by the hot water
$$= 4190 \text{ [J/kg °C]} \times 5 \text{ [kg]} \times (35 - t) \text{ [°C]}$$

Since the loss of heat to external substances is being assumed negligible, the loss of heat by the hot water is equal to the gain of heat by the cold water and vessel; hence,

$$4190 \times 6 \times (t - 8) = 4190 \times 5 \times (35 - t)$$
$$\therefore \qquad\qquad t = 20{\cdot}27\text{°C}.$$

Example 13.5 *A hot copper ball having a mass of 6 kg is lowered into a vessel containing water which, with the vessel, is equivalent to a mass of 5 kg of water at 12°C. The temperature rises to 28°C. Assuming no loss of heat to the surroundings, calculate the initial temperature of the copper ball. Assume the specific heat capacity of copper to be 390 J/kg °C.*

Heat gained by water and vessel
$$= 4190 \text{ [J/kg °C]} \times 5 \text{ [kg]} \times (28 - 12) \text{ [°C]}$$
$$= 335\,200 \text{ J}.$$

If the initial temperature of the copper was t,
heat lost by the copper

$$= 390 \text{ [J/kg °C]} \times 6 \text{ [kg]} \times (t - 28) \text{ [°C]}$$
$$= (2340\ t - 65\ 500) \text{ J}.$$

Since heat lost by copper = heat gained by water and vessel

$\therefore$ $2340\ t - 65\ 500 = 335\ 200$

so that $t = 171°\text{C}.$

13.8 Water equivalent

In the preceding examples, the vessel containing water was taken, with the water, as being together equivalent to a certain mass of water. This is a convenient practice, so it is important to know how the 'equivalent' can be determined. If the vessel is made of metal having a mass m kilograms and a specific heat capacity c joules per kilogram degree Celsius,

heat required to raise its temperature $1°\text{C} = mc$ joules.

If m_W be the water equivalent of the vessel in kilograms, i.e. the mass of water requiring the same amount of heat to raise its temperature $1°\text{C}$: then, from expression (13.2),

$$\left.\begin{array}{l}\text{heat required to raise the temperature} \\ \text{of } m_W \text{ kilograms of water } 1°\text{C}\end{array}\right\} = 4190\ m_W \text{ joules}.$$

Hence, $4190\ m_W = mc$

$\therefore$ water equivalent of vessel = $m_W = mc/4190$ kilograms   (13.4)

For example, if a copper vessel has a mass of 0·8 kg and the copper has a specific heat capacity of 390 J/kg °C,

$$\left.\begin{array}{l}\text{water equivalent} \\ \text{of the vessel}\end{array}\right\} = 0\cdot8 \text{ [kg]} \times 390 \text{ [J/kg °C]}/4190 \text{ [J/kg °C]}$$
$$= 0\cdot0745 \text{ kg}.$$

The vessel and the water content are together equivalent to the mass of water plus the water equivalent of the vessel.

Example 13.6 *An iron bucket has a mass of 2 kg and contains 8 kg of water. Calculate the water equivalent of (a) the bucket alone, (b) the* *bucket and water contained in it, assuming the specific heat capacity of the iron to be 500 J/kg °C.*

If the initial temperature is 6°C, to what temperature will the bucket and water be raised by the addition of 3 kg of water at 95°C?

From expression (13.4),

$$\left.\begin{array}{c}\text{water equivalent of}\\\text{empty bucket}\end{array}\right\} = 2 \text{ [kg]} \times 500 \text{ [J/kg °C]}/4190 \text{ [J/kg °C]}$$
$$= 0{\cdot}238 \text{ kg},$$

∴ water equivalent of bucket and water $= 8 + 0{\cdot}238 = 8{\cdot}238$ kg.

If t be the final temperature after addition of the hot water,

heat gained by bucket and cold water

$$= 4190 \text{ [J/kg °C]} \times 8{\cdot}238 \text{ [kg]} \times (t - 6) \text{ [°C]}$$

and   heat lost by the hot water

$$= 4190 \text{ [J/kg °C]} \times 3 \text{ [kg]} \times (95 - t) \text{ [°C]}.$$

Since this loss of heat is equal to the heat gained by the bucket and cold water,

$$4190 \times 8{\cdot}238 \times (t - 6) = 4190 \times 3 \times (95 - t)$$
∴ $$t = 29{\cdot}8°C.$$

Example 13.7 *A mass of 40 g of aluminium is heated to 200°C and then quickly immersed in 160 g of water contained in a copper vessel having a mass of 24 g, the initial temperature of the water being 12°C. If the final temperature of the water is 21·8°C, calculate the specific heat capacity of the aluminium. Assume the specific heat capacity of copper to be 390 J/kg °C and the loss of heat to be negligible.*

From expression (13.4),

water equivalent of copper vessel

$$= (24/1000) \text{ [kg]} \times 390 \text{ [J/kg °C]}/4190 \text{ [J/kg °C]}$$
$$= 0{\cdot}002\ 23 \text{ kg},$$

∴   water equivalent of vessel and water $= 0{\cdot}16 + 0{\cdot}002\ 23$
$$= 0{\cdot}162\ 23 \text{ kg}.$$

Heat absorbed by vessel and water

$$= 4190 \text{ [J/kg °C]} \times 0{\cdot}162\ 23 \text{ [kg]} \times (21{\cdot}8 - 12) \text{ [°C]}$$
$$= 6660 \text{ J}.$$

If c be the specific heat capacity of the aluminium, then from expression (13.3),

heat given out by the aluminium

$$= c \times (40/1000) \text{ [kg]} \times (200 - 21 \cdot 8) \text{ [°C]}$$
$$= 7 \cdot 128\ c \text{ joules.}$$

$$\left.\begin{array}{l}\text{Since heat given}\\ \text{out by aluminium}\end{array}\right\} = \text{heat absorbed by water and vessel}$$

$$\therefore \qquad\qquad 7 \cdot 128\ c = 6660$$

so that $\qquad\qquad\qquad c = 935 \text{ J/kg °C.}$

13.9 Fuels and calorific values

The most common method of producing heat is the *combustion* or burning of fuels. The energy in fuels is in the form of chemical energy, and burning or combustion is a chemical action in which certain constituents in the fuel, say, coal or oil, unite with the *oxygen* in the atmosphere and, in doing so, give out energy as heat. In this process, the principle of conservation of energy holds good, and the quantity of heat given out is quite definite and can be calculated from a knowledge of the mass and composition of the fuel.

Carbon, which is a constituent of wood, coal, coke and oil, on burning completely gives out about 34 MJ per kilogram. Hydrogen, which is a constituent of coal, coal-gas and oil, releases more than 140 MJ per kilogram. Fuels such as wood, coal, oil and various mixtures of gas vary widely in their composition and in the amount of heat they can give out. A good coal will give out at least 32 MJ per kilogram. The quantity of heat released per kilogram of fuel completely burned is termed the *calorific value* of the fuel. The calorific value of a gas fuel is usually stated in megajoules per cubic metre; thus, the value for 'manufactured' gas is about $18 \cdot 6$ MJ/m^3 and that for natural gas is about $37 \cdot 9$ MJ/m^3.

Example 13.8 *It is required to raise the temperature of 130 litres of water from 15°C to 45°C. How much coal would provide the necessary amount of heat if its calorific value is 32 MJ/kg? How much gas at 18·6 MJ/m^3 would provide the same amount of heat? Neglect any losses.*

Mass of 1 litre of water $= 1$ kg

$\therefore$ mass of 130 litres of water $= 130$ kg.

Increase of temperature $= 45 - 15 = 30°C$.

Hence heat required $= 4190$ [J/kg °C] $\times$ 130 [kg] $\times$ 30 [°C]

$= 16·34 \times 10^6$ J $= 16·34$ MJ.

For coal having a calorific value of 32 MJ/kg,

mass of coal required $= 16·34$ [MJ]/32 [MJ/kg] $= 0·511$ kg.

For gas having a calorific value of 18·6 MJ/m³,

volume of gas required $= 16·34$ [MJ]/18·6 [MJ/m³] $= 0·878$ m³.

Example 13.9 *An engine uses* 11 *kg of oil per hour, the calorific value of the oil being* 44 *MJ/kg. If the engine has an overall efficiency of* 30 *per cent (i.e. it converts* 30 *per cent of the thermal energy into work), calculate the output power of the engine.*

Thermal energy per hour from oil

$= 11$ [kg] $\times$ 44 [MJ/kg] $= 484$ MJ

$= 484 \times 10^6$ J,

$\therefore$ thermal energy per second from oil

$= (484 \times 10^6)/3600$

$= 134\ 400$ J.

Thermal energy converted per second into work

$= 134\ 400$ [J] $\times$ (30/100)

$= 40\ 320$ J,

$\therefore$ output power of engine $= 40\ 320$ W $= 40·32$ kW.

Example 13.10 *A petrol engine developing* 25 *kW drives paddles within a hydraulic brake so that the energy transmitted is converted into heat which warms a stream of water passing through the brake. If the inlet temperature of this stream is* 15°C *and it is desired that the water shall not be heated above* 85°C, *calculate the flow of water that must be provided, in litres per minute, neglecting any heat losses.*

Since 1 kW $= 1$ kJ/s,

work done in 1 second $= 25$ kJ $= 25\ 000$ J.

Rise of temperature of water = 85 − 15 = 70°C.

If m be the mass of water, in kilograms, required per second, then from expression (13.2) we have:

$$25\ 000\ [J] = 4190\ [J/kg\ °C] \times m \times 70\ [°C]$$
$$\therefore \qquad m = 0\cdot0852\ kg.$$

Since the volume of 1 kg of water is practically 1 litre,

∴   volume of water per second = 0·0852 litre

and   volume of water per minute = 0·0852 × 60 = 5·11 litres.

13.10 Change of state: Latent heat

If heat is supplied at a constant rate to a lump of ice having an initial temperature of, say, −20°C, and if no heat is lost by conduction, convection or radiation, the temperature varies as shown by the graph in fig. 13.2. The temperature first increases at a uniform rate from −20°C to 0°C, as shown by line AB. It then remains constant at 0°C for the time BC required for the ice to melt into water. Further supply of heat raises the temperature uniformly to 100°C, as shown by line CD. If the pressure on the surface of the water is atmospheric, the water then begins to boil and the temperature remains constant at 100°C until the water is all evaporated. If the supply of heat is maintained, the temperature of the steam increases, as shown by line EF. The steam is then said to be *superheated*..

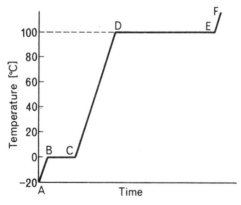

Fig. 13.2 Variation of temperature with time

Another simple experiment to illustrate what happens when a substance changes from the liquid to the solid state is to heat paraffin wax in a test-tube to about 65 or 70°C, i.e. about 10 to 15°C above its melting point. A thermometer is then placed in the liquid wax and the temperature is observed at frequent intervals of time while the wax is cooling. When readings of temperature are plotted against time, the graph is as shown in fig. 13.3, from which it will be seen that during an interval AB, the temperature remains constant. This period represents the time during which the wax is solidifying.

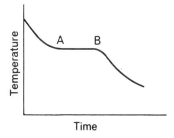

Fig. 13.3 Cooling curve for paraffin wax.

During periods BC and DE in fig. 13.2, when the water is changing from solid to liquid and from liquid to steam respectively, large quantities of heat are absorbed without any change of temperature. Similarly, during period AB in fig. 13.3, the wax loses a relatively large quantity of heat in changing from the liquid to the solid state without any change of temperature.

Let us consider what happens when water changes from solid to liquid and then from liquid to vapour. As mentioned in section 13.1, all materials consist of molecules; and in a solid, these molecules merely vibrate or oscillate about a constant mean position. In a liquid, a molecule can move from one position to another, being hindered only by frequent collisions with other molecules, i.e. they move about with a freedom they do not possess in the solid state. In a vapour, the molecules are much more scattered. There are fewer collisions between molecules though they move about at high speed in random directions. The volume occupied by a given mass of vapour is much greater than that of the same mass of the substance in the solid or liquid form.

The heat absorbed in effecting the change from ice to water and again from water to steam is not apparent as far as indications on a thermometer are concerned. This thermal energy is therefore said to be *latent*. Actually it is absorbed in two ways: (*a*) in breaking the bonds between molecules and (*b*) in the change from water to steam, work is done in pushing on the surrounding atmosphere or other expansible surrounding medium to make room for the increased volume occupied by the steam.

13.11 Specific latent heat of fusion

The heat required to change unit mass of a substance from the solid state to the liquid state at the same temperature is termed the *specific latent heat of fusion* of that substance. For example, when 1 kg of ice at 0°C is melted into water at 0°C, the heat absorbed is 335 kJ; i.e. the specific latent heat of fusion of ice is 335 kJ/kg.

Example 13.11 *Two kilograms of dry ice at 0°C are placed in a vessel containing 5 kg of water at 38°C. The water equivalent of the vessel is 0·2 kg. Calculate the temperature of the water when the ice is completely melted, assuming no loss to or absorption of heat from outside the vessel. Take the specific latent heat of fusion of ice as 335 kJ/kg.*

Let t be the final temperature of the mixture.

Water equivalent of original water and vessel
$$= 5·0 + 0·2 = 5·2 \text{ kg.}$$

From expression (13.2), heat given out by water and vessel
$$= 4190 \text{ [J/kg °C]} \times 5·2 \text{ [kg]} \times (38 - t) \text{ [°C]}$$
$$= (828\ 000 - 21\ 800\ t) \text{ J.}$$

Since specific latent heat of fusion of ice = 335 kJ/kg
$$= 335\ 000 \text{ J/kg}$$

∴ heat required to melt the ice
$$= 2 \text{ [kg]} \times 335\ 000 \text{ [J/kg]} = 670\ 000 \text{ J.}$$

Heat required to raise 2 kg of melted ice from 0°C to t
$$= 4190 \text{ [J/kg °C]} \times 2 \text{ [kg]} \times t$$
$$= 8380\ t \text{ joules.}$$

$$\text{Since heat absorbed in melting} \atop \text{the ice and heating to } t \Biggr\} = \Biggl\{ {\text{heat given out by} \atop \text{water and vessel}}$$

∴ $$670\ 000 + 8380\ t = 828\ 000 - 21\ 800\ t$$

hence $$t = 5 \cdot 24°\text{C}.$$

13.12 Evaporation; Saturation temperature

In water at a constant temperature, some molecules will be moving at a higher velocity than the others; and some of the molecules near the surface of the water may possess sufficient kinetic energy to enable them to escape from the water to form vapour. If the evaporation takes place in an enclosed space, a state of equilibrium is reached when as many molecules return to the liquid in a given time as leave it in that time. The vapour is then said to be *saturated*.

If, however, the evaporation takes place in the open, for example, from a dish containing water, most of the molecules leaving the water do not return to it. If there is a draught blowing over the dish, the vapour is removed and evaporation takes place more rapidly. The heat required to convert the water into vapour comes from the water remaining in the dish; hence the temperature of this water falls. This is the reason why the contents of a jug can be kept cool by wrapping the jug with a wet cloth. The cooling effect is increased by exposing the jug and its wrapping to a draught.

When heat is continuously applied to the liquid, the temperature continues to rise until a value is reached at which the liquid vaporizes with *no further rise of temperature*. The liquid is then boiling and the temperature attained is termed the *boiling point* of the liquid.

The lower the pressure on the surface of the liquid, the lower is the temperature at which the liquid boils. For example, water at standard atmospheric pressure boils at 100°C; whereas at half the standard atmospheric pressure, the boiling point is about 82°C.

The boiling point of water at any particular pressure is also known as the *saturation temperature of steam* at that pressure. The term 'saturation' in this case implies that immediately the steam is subjected to a cooling influence, such as contact with a cooler material, it begins to condense. Steam at the saturation temperature is termed *saturated steam*.

Higher temperatures of steam are possible by the use of *super-heated* or unsaturated steam produced by heating saturated steam which is not in contact with water. Superheated steam does not condense on being subjected to a cooling influence until it has cooled down to the saturation temperature corresponding to its pressure.

13.13. Specific latent heat of vaporization

The heat required to change 1 kg of a substance from the liquid state to vapour at the same temperature is termed the *specific latent heat of vaporization* of that substance. For example, when 1 kg of water at 100°C is converted into steam at 100°C, the heat absorbed is 2257 kJ, i.e. the specific latent heat of vaporization of water at standard atmospheric pressure is 2257 kJ/kg or 2·257 MJ/kg.

The latent heat absorbed in the formation of steam from water is given out in the condensation of steam into water.

Example 13.12 *Calculate the heat required to convert 2 kg of water at 60°C into steam at 100°C at standard atmospheric pressure.*

From expression (13.2),

heat required to raise 2 kg of water from 60°C to 100°C
$$= 4190 \text{ [J/kg °C]} \times 2 \text{ [kg]} \times (100 - 60) \text{ [°C]}$$
$$= 335\,200 \text{ J.}$$

Since specific latent heat of vaporization $= 2257$ kJ/kg
$$= 2·257 \times 10^6 \text{ J/kg,}$$

heat required to convert 2 kg of water at 100°C into steam at 100°C
$$= 2·257 \times 10^6 \text{ [J/kg]} \times 2 \text{ [kg]} = 4·514 \times 10^6 \text{ J.}$$

∴   total heat required $= 0·3352 \times 10^6 + 4·514 \times 10^6$
$$= 4·849 \times 10^6 \text{ J} = 4·849 \text{ MJ.}$$

Example 13.13 *Dry saturated steam (i.e. saturated steam containing no water particles) at 100°C is passed into water contained in a copper calorimeter having a mass of 45 g. The initial mass and temperature of the water are 120 g and 14°C respectively. After the steam has been passed, the mass of the water is 128 g. If the specific latent heat of vaporization of the steam is 2257 kJ/kg, what is the final temperature of the water, assuming the loss of heat to be negligible? The specific heat capacity of copper is 390 J/kg °C.*

From expression (13.4),

water equivalent of calorimeter

$$= 0\text{·}045 \text{ [kg]} \times 390 \text{ [J/kg °C]}/4190 \text{ [J/kg °C]}$$
$$= 0\text{·}0042 \text{ kg.}$$

If t be the final temperature of the water, then from expression (13.2),

heat absorbed by water and calorimeter

$$= 4190 \text{ [J/kg °C]} \times (0\text{·}12 + 0\text{·}0042) \text{ [kg]} \times (t - 14) \text{ [°C]}$$
$$= (520 \, t - 7280) \text{ J.}$$
$$\text{Mass of steam condensed} = 128 - 120 = 8 \text{ g}$$
$$= 0\text{·}008 \text{ kg.}$$

Since specific latent heat of vaporization $= 2257 \text{ kJ/kg}$
$$= 2\text{·}257 \times 10^6 \text{ J/kg,}$$

∴ heat given out by steam in condensing at 100°C

$$= 0\text{·}008 \text{ [kg]} \times 2\text{·}257 \times 10^6 \text{ [J/kg]} = 18\,056 \text{ J,}$$

and heat given out by condensed steam in cooling from 100°C to t

$$= 4190 \text{ [J/kg °C]} \times 0\text{·}008 \text{ [kg]} \times (100 - t) \text{ [°C]}$$
$$= (3350 - 33\text{·}5 \, t) \text{ J.}$$

Equating the heat given out by the steam to that absorbed by the water and calorimeter, we have:

$$18\,056 + 3350 - 33\text{·}5 \, t = 520 \, t - 7280$$
$$\therefore \qquad\qquad t = 51\text{·}8°C.$$

Summary of Chapter 13

The SI unit of temperature is the kelvin (K).
Temperature interval of 1°C = temperature interval of 1 K.

$$t \, (\text{degrees Celsius}) = T \, (\text{kelvins}) - 273\text{·}15 \qquad (13.1)$$

Temperature of melting ice is 273·15 K or 0°C, and temperature of boiling water at standard atmospheric pressure is 373·15 K or 100°C.

Specific heat capacity, c, of a substance is the heat required to raise the temperature of 1 kg of that substance by 1°C and is expressed in joules per kilogram degree Celsius (J/kg °C).

Heat required to raise the temperature of m kilograms of a substance by t degrees $= mct$ joules $\qquad$ (13.3)

$\qquad\qquad\qquad\qquad = 4190\ mt$ joules for water (13.2)

Water equivalent of a vessel $= mc/4190$ $\qquad\qquad$ (13.4)

Specific latent heat of fusion of a substance is the heat required to change 1 kg of the substance from the solid state to the liquid state at the same temperature.

Specific latent heat of fusion of ice $= 335$ kJ/kg.

Specific latent heat of vaporization of a substance is the heat required to change 1 kg of the substance from the liquid state to vapour at the same temperature.

Specific latent heat of vaporization of water at standard atmospheric pressure $= 2257$ kJ/kg.

EXAMPLES 13

1. Express 40°C in the thermodynamic scale and 290 K in the Celsius scale.
2. Calculate the heat, in megajoules, required to raise the temperature of 200 kg of water from 15°C to 90°C.
3. Calculate the heat required to raise the temperature of 30 kg of copper from 12°C to 70°C. Assume the specific heat capacity of copper to be 390 J/kg °C.
4. 90 MJ of heat are absorbed by a body having a mass of 1 Mg to raise its temperature from 20°C to 200°C. Assuming no loss of heat, calculate the specific heat capacity of the body.
5. Calculate the final temperature when 60 kg of water at 80°C are mixed with 200 kg of water at 20°C.
6. When 0·5 kg of hot water is mixed with 2·4 kg of cold water at 8°C, the final temperature is 16°C. Calculate the initial temperature of the hot water. Neglect any loss of heat.
7. A piece of metal having a mass of 0·4 kg is heated to 100°C and then immersed in 0·6 kg of water, at 11·2°C, contained in a calorimeter having a water equivalent of 0·05 kg. The temperature of the water rises to 16·7°C. Calculate the specific heat capacity of the metal, assuming no loss of heat.
8. A brass cylinder, having a mass of 180 g, is heated to 156°C and then immersed in 0·3 litre of water at 14°C. The container has a water equivalent of 16 g. Calculate the temperature attained by the water, assuming no loss of heat. The specific heat capacity of brass is 370 J/kg °C.
9. How much heat is required to raise the temperature of 0·2 kg of aluminium from 18°C to 630°C, assuming the specific heat capacity of aluminium to be 950 J/kg °C?

   If this heated aluminium is immersed in 1·4 kg of water contained in a copper calorimeter at 12°C, calculate the temperature attained by the water, assuming no loss of heat. The mass of the calorimeter is 0·25 kg and the specific heat capacity of copper is 390 J/kg °C.

10. During an experiment in which an iron block was immersed in water, the following readings were obtained: mass of iron, 0·9 kg; initial temperature of iron, 120°C; initial temperature of water, 14°C; final temperature of water and iron, 22°C. Assuming the specific heat capacity of the iron to be 500 J/kg °C and neglecting heat losses and the water equivalent of the container, calculate (*a*) the mass of water in the container and (*b*) the heat, in kilojoules, originally in the iron, measured from 0°C.

11. A piece of steel, having a mass of 0·15 kg, is placed in a hot flue until it attains the same temperature as that of the flue gases. It is then quickly transferred into a copper calorimeter having a mass of 0·4 kg and containing 1 kg of water at a temperature of 15°C. The resultant steady temperature of the steel and water is 22°C. Calculate the temperature of the flue gases, neglecting any losses. Assume the specific heat capacity of steel to be 500 J/kg °C and that of copper to be 390 J/kg °C.

12. In a cooling system, 50 litres of oil per hour are cooled from 120°C to 25°C by means of circulating water. If the water is not to rise in temperature by more than 6°C, calculate the mass of water, in kilograms, required per hour. Assume the oil to have a specific heat capacity of 2100 J/kg °C and its density to be 880 kg/m.^{3}

13. During a brake test on an engine, it was found that 50 per cent of the output power of the engine was absorbed by the water supplied to the brake wheel, the remaining 50 per cent being dissipated as heat to the surrounding atmosphere. The output of the engine was 20 kW, the water flow was 300 litres/hour and the inlet temperature of the water was 15°C. Calculate the final temperature of the cooling water.

14. A bearing absorbs 6 kW in overcoming frictional resistance. Calculate the amount of heat, in kilojoules, produced per minute. If 24 kg of oil are passed through the bearing per minute and absorb 80 per cent of this heat, calculate the rise of temperature of the oil. Assume the specific heat capacity of the oil to be 2000 J/kg °C.

15. When 1·54 g of coal were burnt in a special calorimeter containing 1·63 kg of water, the temperature of the water rose from 14·3°C to 20·4°C. The water equivalent of the metal parts of the calorimeter was 0·23 kg. Calculate the calorific value of the coal in megajoules per kilogram.

16. The temperature of the water flowing through a gas water-heater has to be raised from 15°C to 90°C. The heater uses 90 litres of gas per minute and the heat given out by the gas is 18·6 MJ/m.^{3} Calculate the volume of water, in litres, heated per hour, if 80 per cent of the heat released from the gas passes into the water.

17. If a petrol engine converts 25 per cent of the energy available in the petrol into output energy at the shaft, calculate the output power per kilogram of petrol consumed per hour. Assume the calorific value of the petrol to be 42 MJ/kg.

18. If petrol has a calorific value of 28 MJ/l and if the overall efficiency of a certain petrol engine is 22 per cent when the engine is developing 7·5 kW, calculate the petrol consumption in litres per hour.

19. A lump of dry ice, having a mass of 40 g and an initial temperature of 0°C, is immersed in 600 g of water at 30°C. The container has a water equivalent of 30 g. Calculate the final temperature when all the ice has melted. Assume the latent heat of fusion of ice to be 335 kJ/kg and neglect any loss of heat.
20. A lump of dry ice at 0°C is placed in a calorimeter having a water equivalent of 18 g and containing 300 g of water at 37°C. After the ice has melted, the temperature of the mixture is 14·7°C, and the calorimeter and water are heavier by 76 g. Calculate the specific latent heat of fusion of the ice. Neglect any loss of heat.
21. Calculate the heat, in kilojoules, required to convert 60 g of water at 30°C into saturated steam at 100°C. The container has a water equivalent of 5 g. Assume the specific latent heat of vaporization at 100°C to be 2257 kJ/kg and neglect any loss of heat.
22. Two kilograms of dry saturated steam at 100°C are blown into a vessel containing 120 kg of water at 20°C. Calculate the final temperature of the mixture. Neglect the heat absorbed by the vessel and assume the specific latent heat of vaporization at 100°C to be 2257 kJ/kg.

ANSWERS TO EXAMPLES 13

1. 313·15 K, 16·85°C.	12. 349 kg.
2. 62·85 MJ.	13. 43·6°C.
3. 679 kJ.	14. 360 kJ, 6°C.
4. 500 J/kg °C.	15. 30·8 MJ/kg.
5. 33·8°C.	16. 256 l/h.
6. 54·4°C.	17. 2·92 kW.
7. 450 J/kg °C.	18. 4·39 l/h.
8. 20·8°C.	19. 23·4°C.
9. 116·3 kJ, 31·1°C.	20. 329 kJ/kg.
10. 1·32 kg, 54 kJ.	21. 154·4 kJ.
11. 428°C.	22. 30·2°C.

CHAPTER 14

Expansion of solids, liquids and gases

14.1 Expansion of solids

Most substances increase in volume when they are heated. It has already been mentioned (section 13.3) that the mercury thermometer depends for its action upon the fact that mercury expands more than glass when its temperature is raised. In the case of solid materials, we are usually concerned only with their expansion or contraction in one direction, i.e. with *linear* expansion or contraction. An example of linear expansion and contraction is the shrinking of a steel ring on to a steel shaft. The ring is bored so that its internal diameter, when cold, is slightly less than the diameter of the shaft. The ring is then heated and the increased diameter enables it to be slipped over the shaft; but when the ring cools down, it grips the shaft firmly.

Another example of expansion that has many applications is shown in fig. 14.1, where strips of brass and steel are riveted together.

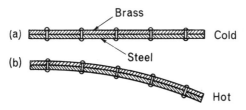

Fig. 14.1 Bimetallic strip.

The expansion of brass is nearly twice that of steel for a given increase of temperature; consequently, when the bimetallic strip is heated, it bends into an arc, as shown in fig. 14.1(*b*). The deflection of the bimetallic strip, for a given increase of temperature, can be practically doubled by using the nickel-steel alloy, *invar*, which contains about 36 per cent nickel, and its coefficient of linear expansion (section 14.2) is about one-sixth of that of ordinary steel. Bimetallic strips are often used in thermostats for controlling the temperature of ovens, etc.

14.2 Coefficient of linear expansion

The amount by which unit length of a material expands when the temperature is raised one degree is termed the *coefficient of linear expansion* of that material and is represented by the Greek letter α (alpha). Hence, if a rod has an initial length l_0 and its temperature is increased by t degrees,

$$\text{increase in length} = \alpha t l_0$$
$$\therefore \quad \text{new length} = l = l_0 + \alpha t l_0$$
$$= l_0(1 + \alpha t) \tag{14.1}$$

The variation in the length of such a rod is represented by the straight line AB of fig. 14.2, where OA represents the length l_0 at the initial temperature and BC represents the length l after the temperature has increased by t degrees.

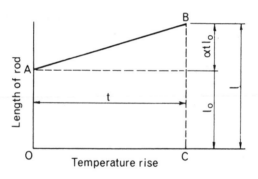

Fig. 14.2 Linear expansion.

The following table gives approximate values of the coefficient of linear expansion for materials that are commonly used in engineering:

Material	Coefficient of linear expansion
Mild steel and wrought iron	$11 \times 10^{-6}/°C$
Copper	$17 \times 10^{-6}/°C$
Aluminium	$23 \times 10^{-6}/°C$
Brass	$20 \times 10^{-6}/°C$

Example 14.1 *A rod is found to be 1·5342 m long at 12°C and 1·5358 m at 100°C. Calculate the value of the coefficient of linear expansion.*

Increase in temperature = $100 - 12 = 88°C$.

Substituting in expression (14.1), we have:

$$1\cdot5358 = 1\cdot5342\,(1\,+\,88\alpha)$$
$$\alpha = 11\cdot85\,\times\,10^{-6}/^{\circ}C.$$

Example 14.2 *The length of a copper wire forming one span of an electric transmission line is 40 m at 10°C. If the coefficient of linear expansion of the copper is* $17\,\times\,10^{-6}/^{\circ}C$, *what is the increase in the length of wire when the temperature rises to 45°C?*

$$\text{Increase in temperature} = 45 - 10 = 35^{\circ}C.$$
$$\text{Initial length of wire} = 40\text{ m}$$
$$\therefore \quad \text{increase in length of wire} = 17 \times 10^{-6}\,[/^{\circ}C] \times 35[^{\circ}C] \times 40[\text{m}]$$
$$= 23\cdot8 \times 10^{-3}\text{ m}$$
$$= 23\cdot8\text{ mm.}$$

14.3 Superficial expansion of a solid

The amount by which unit *area* of a material increases when the temperature is raised by one degree is termed the *coefficient of superficial expansion* and is represented by the Greek letter β (beta). If the initial surface area of the material is a_0 and the temperature is raised by t degrees, the increase in area is $\beta t a_0$, and the new area a is given by:

$$a = a_0 + \beta t a_0$$
$$= a_0(1 + \beta t) \tag{14.2}$$

If we consider a square plate of a material having a coefficient

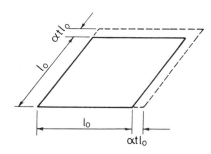

Fig. 14.3 Superficial expansion.

of *linear* expansion α, and if the length of each side of the plate is initially l_0, as shown in fig. 14.3, then $a_0 = l_0^2$. If the temperature is increased t degrees, the increase in length of each side is $\alpha t l_0$

If a is the new area, then:

$$\begin{aligned}
a &= (l_0 + \alpha t l_0)^2 = l_0^2(1 + \alpha t)^2 \\
&= a_0(1 + 2\alpha t + \alpha^2 t^2) \\
&\simeq a_0(1 + 2\alpha t)
\end{aligned} \qquad (14.3)$$

since αt is usually very small compared with unity and $(\alpha t)^2$ is therefore negligible. From a comparison of expressions (14.2) and (14.3), it is seen that

$$\beta \simeq 2\alpha \qquad (14.4)$$

i.e. the coefficient of superficial expansion is approximately twice the coefficient of linear expansion.

14.4 Cubic expansion of solids and liquids

The amount by which unit *volume* of a material increases for one degree rise of temperature is termed the *coefficient of cubic expansion,* and is represented by the Greek letter γ (gamma). Hence, if the initial volume of a material is V_0 and the temperature is increased by t degrees, the increase in volume is $\gamma t V_0$, and the new volume V is given by:

$$\begin{aligned}
V &= V_0 + \gamma t V_0 \\
&= V_0(1 + \gamma t)
\end{aligned} \qquad (14.5)$$

If we consider a cube of solid material having a coefficient of linear expansion α, and if the length of each side is initially l_0, as shown in fig. 14.4, then $V_0 = l_0^3$. For an increase of temperature of t degrees, the increase in the length of each side is $\alpha t l_0$, so that the new volume V is given by:

$$\begin{aligned}
V &= (l_0 + \alpha t l_0)^3 = l_0^3(1 + \alpha t)^3 \\
&= V_0(1 + 3\alpha t + 3\alpha^2 t^2 + \alpha^3 t^3) \\
&\simeq V_0(1 + 3\alpha t)
\end{aligned} \qquad (14.6)$$

since αt is usually very small compared with unity for solids. From a comparison of expressions (14.5) and (14.6), it is seen that

$$\gamma \simeq 3\alpha \qquad (14.7)$$

i.e. the coefficient of cubic expansion of solids is approximately three times the coefficient of linear expansion.

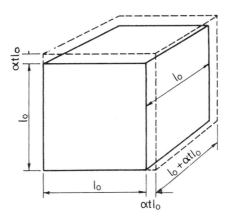

Fig. 14.4 Cubic expansion.

When water cools, contraction occurs until the volume is a minimum at 4°C. Further cooling causes the water to expand until the freezing point is reached. When water is converted into ice, considerable expansion occurs and this often results in the bursting of pipes in very cold weather.

Example 14.3 *A sphere of copper has a diameter of 40 mm. Calculate the increase in volume when the temperature is raised by 180°C. Assume the linear coefficient of expansion of copper to be 17 × 10⁻⁶/°C.*

From expression (14.7),

coefficient of cubic expansion of copper $\simeq 3 \times 17 \times 10^{-6}/°C$
$$= 51 \times 10^{-6}/°C.$$

Initial volume of sphere $= \frac{4}{3}\pi \times (\text{radius})^3$
$$= 1\cdot333 \times 3\cdot14 \times 8000 = 33\,500 \text{ mm}^3$$

Increase in volume $= \gamma t V_0$
$$= 51 \times 10^{-6} \text{ [/°C]} \times 180 \text{ [°C]} \times 33\,500$$
$$\text{[mm}^3\text{]}$$
$$= 308 \text{ mm}^3.$$

14.5 Cubic expansion of gases

For solids and liquids, the value of the coefficient of cubic expansion is so small that no appreciable error is introduced by assuming the initial volume as being that at, or near, the room temperature instead of at 0°C; but in the case of gases, the coefficient of cubic expansion is much larger and it is therefore necessary to be more precise in our definition. Thus *the coefficient of cubic expansion of a gas is the amount by which unit volume of the gas* **at 0°C** *increases when the temperature is increased* 1 *degree, the pressure remaining constant.*

It is found that the volume of many gases such as oxygen, nitrogen and air, at constant pressure, varies as shown by the straight line AB in fig. 14.5, and the slope of this line is such that if the line were extended backwards, it would cut the horizontal axis at a point C

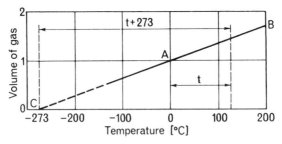

Fig. 14.5 Cubic expansion of a gas at constant pressure.

corresponding to about −273°C. This suggests that if the gas could be cooled to the thermodynamic zero temperature (−273·15°C), its volume would be zero. Actually, the gas would have been first liquefied and then solidified before this temperature would have been reached. From fig. 14.5, it follows that the volume of a gas, at constant pressure, is directly proportional to the thermodynamic temperature or $(t + 273 \cdot 15)$, where t is the temperature in degrees Celsius.

This relationship, first observed by Jacques Charles (1746–1823), a French physicist, is referred to as *Charles' Law* and may be expressed thus: the volume, V, of a given mass of gas, at constant pressure, is directly proportional to the thermodynamic temperature T,[*]

i.e.
$$V = T \times \text{a constant} \tag{14.8}$$

[*] This equation can be used to define temperature, in which case the relationship is arbitrarily postulated and not the result of experiments.

It follows from fig. 14.5 that when a gas remains at constant pressure, the variation of volume for 1°C variation of temperature is $\frac{1}{273}$ of its volume at 0°C,

i.e., coefficient of cubic expansion of a gas at constant pressure
$$= \gamma = (\tfrac{1}{273})/°C.$$

Example 14.4 *A litre of gas at 18°C is heated to a temperature of 150°C, the pressure remaining constant. Calculate the new volume.*

Initial thermodynamic temperature $= 18 + 273 \cdot 15 \simeq 291$ K.

Final thermodynamic temperature $= 150 + 273 \cdot 15 \simeq 423$ K.

From Charles' law,

$$\frac{\text{volume at } 150°C}{\text{volume at } 18°C} = \frac{423 \text{ [K]}}{291 \text{ [K]}}$$

$\therefore$ volume at 150°C $= 1 \text{ [l]} \times 423 \text{ [K]}/291 \text{ [K]}$
$$= 1 \cdot 454 \text{ litres.}$$

14.6 Boyle's law

This law, enunciated by Sir Robert Boyle, an Irish scientist, in 1661, states that *the volume of a given mass of gas at a constant temperature is inversely proportional to the pressure*; thus, if the pressure on a given mass of gas is doubled, the volume is halved. Hence, if V be the volume and p be the pressure, then as long as the gas remains at a constant temperature,

$$V \propto 1/p$$
i.e. $pV = $ a constant   (12.9)

14.7 Combination of Boyle's and Charles' laws

Let us consider a given mass of gas enclosed in a cylinder fitted with a piston, as in fig. 14.6, such that there is no leakage of gas. With the piston in the position shown in fig. 14.6(*a*), let us assume the absolute pressure, the volume and the thermodynamic temperature to be represented by p_1, V_1 and T_1 respectively.

Suppose that the three quantities—pressure, volume and temperature—vary simultaneously to p_2, V_2 and T_2 respectively, and that we require to know the value of one of the quantities when the new values of the other two quantities are known.

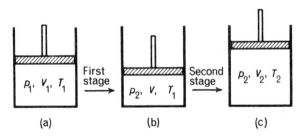

Fig. 14.6 Boyle's and Charles' laws.

The relationship between the three quantities can be derived by imagining the change to take place in two stages in such a way that Boyle's law can be applied to the first stage and Charles' law to the second stage, thus:

(*a*) if the pressure is varied from p_1 to p_2, with the temperature remaining constant, the volume changes from V_1 to a value V, fig. 14.6(*b*); then by Boyle's law,

$$p_1 V_1 = p_2 V$$

$$\therefore \qquad V = \frac{p_1 V_1}{p_2} \qquad (14.10)$$

(*b*) if the temperature is then varied from T_1 to T_2, the pressure remaining constant, the volume changes from V to V_2, fig. 14.6(*c*); then by Charles' law,

$$\frac{V}{T_1} = \frac{V_2}{T_2}$$

$$\therefore \qquad V = \frac{V_2 T_1}{T_2} \qquad (14.11)$$

Equating expressions (14.10) and (14.11), we have:

$$\frac{p_1 V_1}{p_2} = \frac{V_2 T_1}{T_2}$$

so that

$$\frac{p_1 V_1}{T_1} = \frac{p_2 V_2}{T_2} = \frac{p_3 V_3}{T_3} \text{ etc.}$$

Hence, in general,

$$\frac{pV}{T} = \text{a constant} \qquad (14.12)$$

This relationship is very useful for converting the volume of a gas determined experimentally at convenient temperature and pressure to the volume at standard temperature of 0°C and standard atmospheric pressure of 101·325 kPa (or 760 mm of mercury). Such a condition is referred to as *standard temperature and pressure*, abbreviated to *s.t.p.*

In all problems involving gas laws, it is the *absolute pressure* and not the *gauge pressure* that must be used. The gauge pressure—as the name implies—is the reading on a pressure gauge and represents the amount by which the actual pressure differs from the atmospheric pressure. Thus, if a gauge attached to a steam pipe reads 1200 kPa and if the actual atmospheric pressure outside the pipe is 100 kPa, the absolute pressure of the steam is 1300 kPa. In general,

absolute pressure = gauge pressure + atmospheric pressure.

Example 14.5 *Air at 8°C and a gauge pressure of 300 kPa is passed through a heater. If the temperature is raised to 50°C and the gauge pressure falls to 230 kPa, what is the percentage change in volume? Assume the atmospheric pressure to be 101·3 kPa.*

Initial absolute pressure = 300 + 101·3 = 401·3 kPa.
Final absolute pressure = 230 + 101·3 = 331·3 kPa.
Initial temperature = 8 + 273·15 ≃ 281 K
Final temperature = 50 + 273·15 ≃ 323 K.

Suppose the initial volume of the air to be 100 m³, then if V be the final volume in cubic metres, we have from expression (14.12),

$$\frac{401\cdot3 \,[\text{kPa}] \times 100\,[\text{m}^3]}{281\,[\text{K}]} = \frac{331\cdot3\,[\text{kPa}] \times V}{323\,[\text{K}]}$$

∴ $V = 139$ m³.

Hence,   increase in volume = 39 per cent.

Example 14.6 *A quantity of gas at a pressure of 1150 mm of mercury and a temperature of 75°C occupies a volume of 2·4 l. Calculate the volume at standard temperature and pressure. Standard pressure corresponds to a pressure of 760 mm of mercury.*

Initial temperature of gas $\simeq 75 + 273 = 348$ K.

Let V be the volume of the gas, in litres, at standard temperature and pressure. Substituting in expression (14.12), we have:

$$\frac{1150 \text{ [mmHg]} \times 2\cdot4 \text{ [l]}}{348 \text{ [K]}} = \frac{760 \text{ [mmHg]} \times V}{273 \text{ [K]}}$$

∴ $V = 2\cdot85$ litres.

Example 14.7 *Calculate the mass of dry air in a room, $8\,m \times 6\,m \times 4\,m$, when the atmospheric pressure is $96\,kPa$ and the temperature is $20°C$. The mass of $1\,m^3$ of dry air at s.t.p. is $1\cdot29\,kg$.*

Volume of room $= 8 \times 6 \times 4 = 192$ m³.

Temperature of room $= 20°C$

$\simeq 20 + 273 = 293$ K.

Standard atmospheric pressure $\simeq 101\cdot3$ kPa.

If V be the volume of the air, in cubic metres, at s.t.p., then:

$$\frac{96 \text{ [kPa]} \times 192 \text{ [m}^3\text{]}}{293 \text{ [K]}} = \frac{101\cdot3 \text{ [kPa]} \times V}{273 \text{ [K]}}$$

∴ $V = 169\cdot7$ m³

and   mass of air $= 169\cdot7 \times 1\cdot29 = 219$ kg.

14.8 Ideal gases

An ideal gas can be defined as being one that obeys the equation

$pV/T =$ a constant

The value of the constant depends on the gas considered and its mass. If 1 kg of gas is considered

$pV/T = R$ (14.13)

where R is called the specific gas constant. The following are some typical values.

Gas	$R\,[\mathrm{N}\,m\,\mathrm{kg}^{-1}\,\mathrm{K}^{-1}]$
Dry air	287
Oxygen	260
Nitrogen	297
Steam at low pressure	462

Superheated steam can be considered to approximate to an ideal gas, particularly at low pressures.

If m kg of a gas is considered then the equation becomes

$$pV/T = mR \qquad (14.14)$$

The specific gas constant is just multiplied by the mass, in kg, concerned to give the constant for the ideal gas equation.

Example 14.8 *Using the value of the characteristic gas constant given in the above table, calculate the mass of* $1 \cdot 5\,m^3$ *of dry air at* $20°C$ *when under a pressure of* 4×10^5 *Pa.*

Using equation (14.14) and rearranging it

$$m = \frac{pV}{TR}$$

$$= \frac{4 \times 10^5\,[\mathrm{Pa}] \times 1 \cdot 5\,[m^3]}{293\,[\mathrm{K}] \times 287\,[\mathrm{N}\,m\,\mathrm{kg}^{-1}\,\mathrm{K}^{-1}]}$$

$$= 7 \cdot 1\,\mathrm{kg}$$

Summary of Chapter 14

Coefficient of linear expansion of a solid is the amount by which unit length increases for one degree rise of temperature.

$$l = l_0(1 + \alpha t) \qquad (14.1)$$

Coefficient of superficial expansion of a solid is the amount by which unit area increases for one degree rise of temperature.

$$a = a_0(1 + \beta t) \qquad (14.2)$$

Coefficient of cubic expansion of a substance is the amount by which unit volume increases for one degree rise of temperature.

$$V = V_0(1 + \gamma t) \qquad (14.5)$$

In the case of a gas, V_0 in expression (14.5) is the volume at 0°C,

t is the temperature above 0°C and the pressure is assumed to remain constant.

For solids: $\beta \simeq 2\alpha$ (14.4)

and $\gamma \simeq 3\alpha$ (14.7)

For a gas: $\gamma = (\frac{1}{273})/°C$.

Charles' law. The volume of a given mass of gas, at constant pressure, is directly proportional to the thermodynamic temperature,

i.e. $V/T = $ a constant   (14.8)

Boyle's law. The volume of a given mass of gas, at constant temperature, is inversely proportional to the absolute pressure,

i.e. $pV = $ a constant   (14.9)

For a given mass of gas, the product of the absolute pressure and the volume divided by the thermodynamic temperature is a constant,

i.e. $pV/T = $ a constant   (14.12)

A gas which obeys this relationship is known as an ideal gas.

For 1 kg of gas

$$pV/T = R$$

EXAMPLES 14

1. A metal rod is 500 mm long at 30°C and 500·5 mm at 130°C. Calculate the coefficient of linear expansion of the metal.
2. If a brass scale is correct length at 5°C, calculate the percentage increase in length at 60°C. The coefficient of linear expansion of brass is 0·000 02/°C.
3. A steel pipe is 20 m long at 12°C. Calculate the increase in length when steam at 130°C is passing through it, assuming the coefficient of linear expansion of the steel to be 0·000 011/°C.
4. In an experiment, it was found that a copper rod, 400 mm long at the initial temperature, increased in length by 0·544 mm when the temperature was raised by 80°C. Calculate the coefficient of linear expansion of the copper.
5. The length of an aluminium conductor forming one span between the towers of an electric transmission line is 120 m. What is the variation in the length of the conductor when the temperature varies between −20°C and 55°C? Assume $\alpha = 0·000\ 023/°C$ for the aluminium.

6. A steel wire, 2 mm in diameter, is stretched without appreciable tension between two fixed points at a temperature of 25°C. Calculate the tension when the temperature falls to 0°C. Assume $\alpha = 0.000\ 011/°C$ and $E = 220$ GPa.

7. A brass plate is 400 mm square at 12°C. If the temperature is raised to 100°C, what is the increase in the superficial area of the plate? Assume the coefficient of linear expansion of brass to be $0.000\ 02/°C$.

8. A block of cast iron measures 24 mm $\times$ 15 mm $\times$ 10 mm when placed in melting ice. What is the increase in volume when the block is placed in steam at standard atmospheric pressure, if the coefficient of linear expansion of cast iron is $0.000\ 01/°C$?

9. The mercury contained in a certain thermometer has a volume of 540 mm^3 at 5°C. What is the increase in volume when the temperature is raised to 85°C? Assume the coefficient of cubic expansion of mercury to be $0.000\ 18/°C$.

10. Ethyl alcohol has a coefficient of cubic expansion of $0.0011/°C$. Calculate the reduction in volume when the temperature of 0.05 litre of ethyl alcohol is reduced from 40°C to −15°C.

11. A volume of 0.4 litre of dry air at 10°C is heated to a temperature of 120°C, the pressure remaining constant. Calculate the new volume.

12. A quantity of gas at constant pressure and a temperature of 27°C is heated to 60°C and becomes 0.5 m^3. What was the original volume?

13. If the pressure on a gas remains constant and its initial volume and temperature are 1 m^3 and 12°C respectively, what will be (a) its volume if the temperature is raised to 70°C and (b) the temperature when its volume increases to 1.3 m^3?

14. A certain gas has a volume of 50 m^3 at 200°C. If the pressure is maintained constant while the temperature is reduced to −50°C, what is the final volume? What is the thermodynamic temperature of the gas when the volume is reduced to 40 m^3?

15. The density of air at s.t.p. is 1.29 g/l. Calculate the mass of air contained in a flask of volume 2.5 litres at standard pressure when the temperature is (a) 80°C and (b) 230 K.

16. Four cubic metres of gas are compressed from a pressure of 100 kPa to a pressure of 500 kPa, the temperature remaining constant. What is the final volume?

17. Air is compressed in a cylinder from a pressure of 105 kPa and a temperature of 20°C to one-quarter of its original volume at constant temperature. What is the resulting pressure? The air is then heated at constant pressure until it occupies its original volume. Calculate the final temperature of the air in degrees Celsius.

18. A certain mass of air has a volume of 0.56 litre at a pressure equivalent to 780 mm of mercury and a temperature of 5°C. What is the volume when the pressure is equivalent to 710 mm of mercury and the temperature is 25°C?

19. An oxygen cylinder contains 110 litres of gas at an absolute pressure of 1000 kPa and a temperature of 12°C. What would be the volume of the

gas at standard atmospheric pressure of 101·3 kPa and a temperature of 28°C?

20. Compressed air at a pressure of 400 kPa and a temperature of 15°C is passed through a heater. If the temperature is raised to 70°C and the pressure falls to 300 kPa, what is the percentage change in volume?

21. A gas has a volume of 0·8 litre at 30°C and a pressure of 90 kPa. Calculate the volume at s.t.p.

22. A volume of gas occupying 0·8 litre at s.t.p. is heated until its pressure is 115 kPa. If the final volume is 0·92 litre, what is the temperature of the gas?

23. A compressed-air cylinder is 2 m long internally and has an internal diameter of 0·6 m. It contains air at a pressure of 5 MPa and a temperature of 15°C. If 1 m³ of air at s.t.p. has a mass of 1·29 kg, calculate the mass of air in the cylinder.

ANSWERS TO EXAMPLES 14

1. $10 \times 10^{-6}/°C$.
2. 0·11 per cent.
3. 25·96 mm.
4. $17 \times 10^{-6}/°C$.
5. 207 mm.
6. 190 N.
7. 563 mm².
8. 10·8 mm³.
9. 7·78 mm³.
10. 3025 mm³.
11. 0·555 litre.
12. 0·45 m³.
13. 1·203 m³, 97·5°C.
14. 23·6 m³, 378·6 K.
15. 2·49 g, 3·825 g.
16. 0·8 m³.
17. 420 kPa, 900°C.
18. 0·66 litre.
19. 1145 litres.
20. 59 per cent increase.
21. 0·64 litre.
22. 83°C.
23. 34·2 kg.

CHAPTER 15

Electric charges

15.1 Electrification by friction

The fact that amber, when rubbed, acquires the property of attracting light bodies was referred to over 2500 years ago by a Greek philosopher, named Thales, as a phenomenon that was quite familiar at that time. It was not until about A.D. 1600 that Dr Gilbert of Colchester discovered that other bodies, such as glass, could also be electrified by rubbing.

If a glass rod is rubbed with silk and placed on a stirrup hung from a wooden stand (fig. 15.1), and if the end of another similarly rubbed glass rod is brought near one end of the suspended rod, they are found to repel each other. If an ebonite rod rubbed with fur is brought near the suspended glass rod, attraction takes place. Similarly, if the charged ebonite rod is supported on the stirrup, the ·approach of another charged ebonite rod causes repulsion, whereas a charged glass rod causes attraction.*

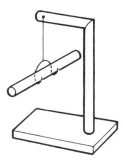

Fig. 15.1 Attraction and repulsion of electric charges.

The glass and the ebonite appear to be charged with different kinds of electricity; and the above experiments show that bodies

---

* These experiments are difficult to perform satisfactorily if the atmosphere is humid, but they do form a very simple and useful introduction to the idea of positive and negative charges of electricity and to the existence of forces of repulsion between like charges and of attraction between unlike charges.

H

charged with the same kind of electricity repel, while bodies charged with opposite kinds of electricity attract one another.

About 1750, Benjamin Franklin suggested that electricity was some form of fluid which passed from one body to the other when they were rubbed together, and that in the case of glass rubbed with silk, the electric fluid passed from the silk into the glass so that the glass contained a 'plus' or 'positive' amount of electricity. On the other hand, when ebonite was rubbed with fur, the electric fluid passed from the ebonite to the fur, leaving the ebonite with a 'minus' or 'negative' amount of electricity. Franklin's one-fluid theory has long been discarded, but his convention still remains: thus, the glass is said to be positively charged and the ebonite negatively charged, and it is this convention which governs the signs used for the terminals of batteries and direct-current generators.

We may now summarise the results of the above experiments by stating that like charges repel and unlike charges attract each other, as shown in fig. 15.2, where the arrows represent the directions of the forces on the charges.

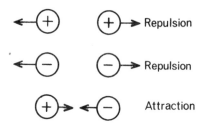

Fig. 15.2 Repulsion and attraction between electric charges.

15.2 Structure of the atom

It was mentioned in section 13.1 that all substances consist of molecules, the molecule being the smallest particle of a substance that can have a separate existence and still retain the characteristics of that substance. A molecule consists of one or more particles called *atoms*, an atom being the smallest particle of an element which can take part in a chemical reaction. Molecules of helium, neon and argon contain one atom each and are therefore termed *monatomic* molecules. Molecules of hydrogen, oxygen and nitrogen consist of two atoms each and are therefore said to be *diatomic*. A molecule of sulphur consists of eight atoms.

Every atom consists of a relatively massive core or nucleus carrying a positive charge, around which *electrons* can be thought of as moving in orbits at distances that are great compared with the size of the nucleus. Each *electron* has a mass of $9 \cdot 11 \times 10^{-31}$ kg and a *negative* charge, $-e$. The nucleus of every atom except that of hydrogen consists of *protons* and *neutrons*. Each *proton* carries a *positive* charge, **e**, equal in magnitude to that of an electron and its mass is $1 \cdot 673 \times 10^{-27}$ kg, namely 1836 times that of an electron. A *neutron*, on the other hand, carries *no* resultant charge and its mass is approximately the same as that of a proton. Under normal conditions, an atom is neutral, i.e. the total negative charge on its electrons is equal to the total positive charge on the protons.

The atom possessing the simplest structure is that of hydrogen. It consists merely of a nucleus of one proton together with a single electron which may be thought of as revolving in an orbit, of about 10^{-7} mm diameter, around the proton, as in fig. 15.3(a).

Fig. 15.3(b) shows the arrangement of a helium atom. In this case, the nucleus consists of two protons and two neutrons, with two electrons orbiting in what is termed the *K shell*. This shell is complete with only two electrons. Consequently the helium atom is stable and does not combine with any other atom, i.e. helium is an *inert* element.

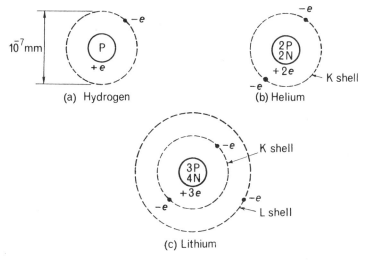

Fig. 15.3 Hydrogen, helium and lithium atoms.

The element with *three* orbital electrons is lithium. Two of these electrons form the K shell, as before, and the third electron starts another shell, known as the L shell, as shown in fig. 15.3(c).

The number of electrons in the L shell of the elements beryllium, boron, carbon, nitrogen, oxygen, fluorine and neon are 2, 3, 4, 5, 6, 7 and 8 respectively. With eight electrons, the L shell is complete; hence the reason why neon is another inert element.

Further increase in the number of orbital electrons results in the formation of a third shell, referred to as the M shell. For instance, the element sodium has eleven orbital electrons, two of which form the K shell, eight the L shell and the remaining electron forms the M shell.

The electrons orbiting in the outermost shell are termed *valence electrons*. The number of valence electrons determines the properties of an element; for instance, the elements copper, silver and gold, which are good conductors of electricity, have a single electron orbiting in the N, O and P shells respectively.

An atom which has lost or gained one or more electrons is referred to as an *ion*; thus, for an atom which has lost one or more electrons, its negative charge is less than its positive charge and such an atom is therefore termed a *positive ion*.

15.3 Movement of free electrons in a metal

When atoms are packed tightly together, as in a metal, each outer electron experiences a small force of attraction towards neighbouring nuclei, with the result that such an electron is no longer bound to any individual atom, but can move at random within the metal. These electrons are termed *free* or *conduction* electrons, and only a slight external influence is required to cause them to drift in a desired direction.

The full lines AB, BC, CD, etc. in fig. 15.4 represent paths of the random movement of one of the free electrons in a *metal* rod when there is no cell connected across the terminals EF; i.e. the electron is accelerated in direction AB until it collides with an atom with the result that it may rebound in direction BC, etc. Different free electrons move in different directions so as to maintain the electron density constant throughout the metal; in other words, there is no resultant drift of electrons towards either E or F.

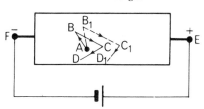

Fig. 15.4 Movement of a free electron.

If a cell, such as that used in an electric torch, is connected across terminals E and F, the cell terminal marked '+' being connected to terminal E, the latter becomes positive relative to F. The effect is to modify the random movement of the electron as shown by the dotted lines AB_1, B_1C_1 and C_1D_1 in fig. 15.4, i.e. there is superimposed on the random movement a drift of the electron towards the positive terminal E. The number of electrons reaching terminal E from the rod is the same as that entering the rod at terminal F. It is this drift of the electrons that constitutes an electric current in the circuit.

In materials such as glass and porcelain at normal temperature, the electrons are tightly bound to their respective nuclei so that there are very few free electrons present. Consequently such materials are very poor conductors of electric charges and are therefore termed *insulators*.

15.4 Current and electron movement

It was mentioned In section 15.1 that the convention governing the marking of the terminals of a battery as 'positive' and 'negative' is based upon a theory propounded over 200 years ago by Benjamin Franklin. This theory suggested that the positive terminal of a battery is at a *higher electrical potential* than the negative terminal. Consequently, when a metal wire is connected across the terminals, electricity is assumed to flow through the wire *from* the positive terminal *to* the negative terminal. Unfortunately, this theory has proved incorrect. We now know that the electric current in a metal wire connected across a battery consists of the movement of electrons from the negative terminal of the battery to the positive terminal, i.e. from a point at the *lower* potential to a point at the *higher* potential, as indicated in fig. 15.5.

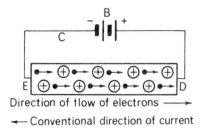

Direction of flow of electrons ⟶

⟵ Conventional direction of current

Fig. 15.5 Movement of free electrons in a metal.

The convention for the direction of an electric current, based on Franklin's theory, had been universally adopted long before the discovery of the electron in 1897, and so we continue to say that an electric current flows from a point at the *higher* potential to that at the *lower* potential, as shown in fig. 15.5.

15.5 Units of quantity of electricity and of electric current

When a current is flowing in a circuit, suppose n to be the number of electrons passing a given point of the circuit in a given time. Since each electron carries a negative charge, $-e$, it follows that the total electric charge that has passed the given point is $-ne$. This total charge is referred to as the *quantity of electricity* that has passed that point.

The unit of quantity of electricity is the *coulomb* (symbol, C), named after a French physicist, C. A. de Coulomb (1736–1806). The magnitude of the coulomb is enormous compared with that of the charge on an electron; in fact, the coulomb is 6·24 million million million (6·24 × 10^{18}) times the charge on an electron.

Expressed in another way, the charge on an electron

$$= \frac{1}{6\cdot24 \times 10^{18}} = 1\cdot602 \times 10^{-19} \text{ coulomb.}$$

The rate at which electricity flows past any given point of an electric circuit is termed an *electric current*. The unit of current is the *ampere* (symbol, A), named after another French scientist, André-Marie Ampère (1775–1836), and is one of the seven SI base units listed in section 1.1. We shall defer giving the definition of the ampere until a later stage. All we need say at this stage is that if I represents the value of the current, in amperes, flowing in a circuit for time *t*, in seconds, then the quantity of electricity *Q*, in coulombs, that has passed any given point of the circuit in that time is:

$$Q = I \times t$$
or
$$I = Q/t \qquad (15.1)$$

Example 15.1 *The current through a certain lamp is 0·6 A. If this current remains constant for 2 hours, what is the quantity of electricity, in coulombs, that flows through the lamp during that time?*

Time $= 2$ [h] $\times$ 3600 [s/h] $= 7200$ s,

$\therefore$ quantity of electricity $= 0·6$ [A] $\times$ 7200 [s]
$$= 4320 \text{ C.}$$

Example 15.2 *If the current in a circuit is 1 A, calculate the number of electrons passing any given point of the circuit per second.*

Quantity of electricity per second $= 1$ [A] $\times$ 1 [s]
$$= 1 \text{ C.}$$

Since each electron carries a negative charge of $1·602 \times 10^{-19}$ C,

$\therefore$ number of electrons passing a given point of the circuit per second

$$= \frac{1}{1·602 \times 10^{-19}} = 6·24 \times 10^{18},$$

i:e. when the current in a circuit is 1 ampere, electrons are passing any given point of the circuit at the rate of $6·24 \times 10^{18}$ per second.

Summary of Chapter 15

Matter consists of an aggregate of atoms, each of which has a nucleus of protons and neutrons surrounded by electrons which can be considered as moving in orbits around the nucleus. The protons carry positive charges and the electrons negative charges, but the neutrons carry no resultant charge. When the atom is in its normal or neutral state, its positive and negative charges are equal in magnitude. In metals, one or more of the outermost electrons are loosely held to the atom and can be made to drift from one atom to another in a desired direction by connecting an electric cell across the circuit.

The unit of quantity of electricity is the coulomb and that of current is the ampere.

The current is the rate at which electricity flows in a circuit,

i.e. $$I = Q/t \qquad (15.1)$$

An electron has a negative charge of 1.602×10^{-19} C, and when the current in a circuit is 1 A, the number of electrons passing a given point of the circuit is 6.24×10^{18} per second.

EXAMPLES 15

1. The current through an electric heater is 4 A. If this current remains constant for 3 hours, calculate the quantity of electricity, in coulombs, that has passed a given point of the circuit.
2. If the quantity of electricity that passes a given point of a circuit in 40 s is 600 C, what is the average value of the current?
3. If the number of electrons flowing per second through a circuit is 6×10^{16}, what is the value of the current in milliamperes?
4. The current in a circuit is 30 μA. Calculate the rate at which electrons pass any given point of the circuit.
5. If there are 3×10^{14} electrons passing per second across the space between two metal surfaces, calculate (*a*) the current, in microamperes and (*b*) the quantity of electricity, in microcoulombs, that passes in 5 s.

ANSWERS TO EXAMPLES 15

1. 43 200 C.   4. 187.2×10^{12} electrons/second.
2. 15 A.   5. 48 μA, 240 μC.
3. 9.6 mA.

CHAPTER 16

Electric current

16.1 Effects of an electric current

All the phenomena which an electric current may produce can be grouped under the following headings:

(*a*) magnetic effect, (*b*) heating effect, (*c*) chemical effect.

These three effects can be demonstrated by connecting the following items to a direct-current supply:

(i) a coil C wound on an iron core bent into a U shape (fig. 16.1a),
(ii) an electric fire element or an electric lamp,

(iii) a double-pole change-over switch D, to which are also connected a flash-lamp bulb L and a pair of lead plates, P and Q, dipping into a dilute solution of sulphuric acid in water (fig. 16.1b).

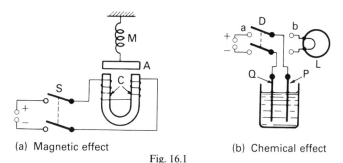

(a) Magnetic effect   (b) Chemical effect

Fig. 16.1

If an iron plate or armature A is suspended by a spring M a little above the ends of the iron core, it will be attracted downwards immediately S is closed and a current flows through the coil. In other words, the iron core becomes magnetized and mechanical work is done in extending M.

An electric fire element or the filament of a light bulb is found to become warm and its temperature may even rise sufficiently for the wire to glow, thereby giving out light energy as well as thermal energy, when current flows through them.

After switch D has been over to side *a* and the current has been flowing for a few minutes, let us open S and switch D over to side *b*. The filament of lamp L becomes incandescent, but its brightness gradually fades away. If the plates P and Q are withdrawn from the solution, P is found to have a faint chocolate-colour coating while Q appears unaltered. In this case, electrical energy has been converted into chemical energy in changing the lead on the surface of plate P into an oxide of lead. This chemical action happens to be of a kind that is reversible. Consequently, when the plates are connected to the lamp, chemical energy is converted back into electrical energy which is then converted into thermal and light energies. This reversible chemical action is the basis upon which secondary batteries operate (Chapter 21).

16.2 Magnetic effect

In fig. 16.2, E and F represent the cross-section of two co-axial coils, E being suspended from a spring balance G about 2 cm above F which rests on a table or bench. The coils are connected so that they carry current in the *same* direction, the current being led into and out of coil E by flexible wires. Another coil, B, also shown in section in fig. 16.2, is wound on a hollow cylindrical former and an iron core C is suspended from a spring balance S.

The three coils, E, F and B, are connected in series with a variable resistor R, and a switch N across a direct-current supply.

With switch N closed, an electric current flows in the circuit. The magnetic effect produced by the current in coils E and F causes them to attract each other (the reason for this behaviour is given in Chapter 19). Also, the iron core C is attracted towards coil B. The greater the current, the larger are the readings on balances G and S, and marks can be inserted on the two scales to register the various currents indicated by ammeter A. Whenever a current flows a magnetic effect is produced.

16.3 Chemical effect

The glass vessel B in fig. 16.3 contains a solution of copper sulphate in water. Two copper plates, P and Q, are partially immersed in the *electrolyte* (an electrolyte being any liquid that can be decomposed electrically) and are connected in series with an ammeter A, a variable resistor R and a switch N to a direct-current supply.

Plate Q, namely the plate connected to the negative terminal of the

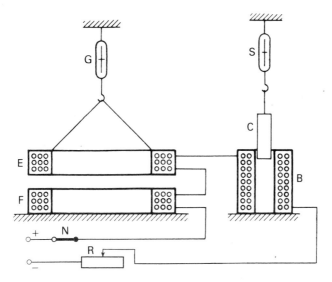

Fig. 16.2 Use of the magnetic effect to measure an electric current.

supply, is washed, dried and weighed before it is placed in the copper sulphate solution. Switch N is then closed and the current is maintained constant at a known value for, say 10 minutes. Switch N is then opened and plate Q is removed, washed, dried and again weighed. It is found that the mass of the plate has increased and the exact amount is determined from the difference between the balance readings.

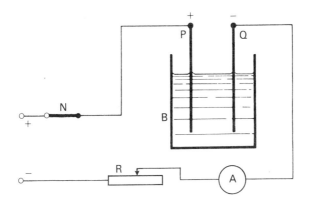

Fig. 16.3 Use of the chemical effect to measure an electric current.

In this way we can measure accurately the amount of copper deposited by a *given current* in, say, 10, 20 and 30 minutes, and also the amount deposited in a *given time* by different values of the current. It is found that the mass of copper deposited is directly proportional to the current and to the time; in other words, it is proportional to the *quantity* of electricity.

Owing to the accuracy and the ease with which the increase in the mass of a plate can be determined, this chemical effect of an electric current may be employed for measuring the quantity of electricity with a high degree of precision. At one time, this was the principle of the method used for defining the magnitude of the unit of current.

16.4 The ampere and the coulomb

It was stated in section 15.5 that the SI unit of current is the *ampere*, that the unit of quantity of electricity is the *coulomb* and that if a current I, in amperes, is maintained for time t, in seconds, the corresponding quantity of electricity, in coulombs, is represented by Q, where

$$Q = I \times t$$

or

$$I = Q/t \qquad (15.1)$$

In 1948 it was decided internationally that the *ampere* should be defined as *the constant current which, if maintained in two straight parallel conductors of infinite length, of negligible circular cross-section and placed at a distance of* 1 *metre apart in a vacuum, would produce between them a force equal to* 2×10^{-7} *newton per metre length*. The reason for the existence of a force between two parallel current-carrying conductors is given in Chapter 19.

The apparatus required to measure a current accurately in terms of the above definition is based upon the principle of the attraction between two current-carrying coils, E and F, of fig. 16.2; but its construction is so elaborate and expensive that it is only available at the principal national laboratories, such as the National Physical Laboratory in this country.

From carefully-conducted experiments based upon the chemical effect of an electric current described in section 16.2, it has been found that for every coulomb of electricity passing between two copper plates immersed in a copper sulphate solution, 0·3294 mg of copper is deposited on the negative plate (plate Q in fig. 16.3). An arrangement such as that shown in fig. 16.3, when used to measure the quantity of electricity by the amount of a substance liberated electro-chemically, is known as a *coulometer* or *voltameter*.

Previous to 1948, the ampere was defined in terms of the chemical effect produced when current was passed through a silver coulometer consisting of two silver plates immersed in a silver nitrate solution. The silver coulometer can still be used as a sub-standard for measuring the value of the current on the assumption that 1 coulomb of electricity deposits 1·1182 mg of silver on the negative plate.

Since 1 ampere is 1 coulomb/second, the unit quantity of electricity may also be termed an *ampere second*. For many purposes, such as for stating the quantity of electricity a battery is capable of supplying, the coulomb is inconveniently small, and a larger unit, known as the *ampere hour* (A h), is preferable. Thus, if a battery supplies 4 A for 10 h, the quantity of electricity

$$= 4 \text{ [A]} \times 10 \text{ [h]} = 40 \text{ A h}.$$

Example 16.1 *The current through a silver coulometer is maintained constant for 15 min. The initial and final masses of the negative plate are 16·347 g and 17·518 g respectively. Calculate the value of the current, assuming that 1 C of electricity deposits 1·1182 mg of silver on the negative plate.*

Mass of silver deposited $= 17·518 - 16·347 = 1·171$ g
$$= 1171 \text{ mg}.$$

$\therefore$ quantity of electricity $\}$ $= \dfrac{1171 \text{ [mg]}}{1·1182 \text{ [mg/C]}}$
through coulometer $\int$

$$= 1047·2 \text{ C}.$$

Duration of test $= 15$ min
$$= 15 \times 60 = 900 \text{ s},$$

$\therefore$ current $= \dfrac{1047·2 \text{ [C]}}{900 \text{ [s]}} = 1·163 \text{ A}.$

Example 16.2 *If a current of 15 A is maintained constant for 20 min, calculate the quantity of electricity in* (a) *coulombs and* (b) *ampere hours.*

(*a*) Quantity of electricity, in coulombs

$$= 15 \text{ [A]} \times (20 \times 60) \text{ [s]} = 18\ 000 \text{ C.}$$

(*b*) Quantity of electricity, in ampere hours

$$= 15 \text{ [A]} \times (20/60) \text{ [h]} = 5 \text{ A h.}$$

16.5 Electrochemical equivalent

In section 16.2 it was found that the mass of copper deposited on the negative plate (or *cathode*) was proportional to the quantity of electricity passing through the electrolyte. This relationship was discovered by Michael Faraday in 1832 when he enunciated two laws:

(*a*) the amount of chemical change produced by an electric current is proportional to the quantity of electricity,

(*b*) the amounts of different substances liberated by a given quantity of electricity are proportional to their chemical equivalent mass, where

$$\text{chemical equivalent mass} = \frac{\text{relative atomic mass}}{\text{valency}}$$

The relative atomic masses and valencies of some of the most common elements are given in the following table.

The mass of a substance liberated from an electrolyte by 1 coulomb is termed the *electrochemical equivalent* of that substance; thus, the electrochemical equivalents of copper and silver are respectively 0·3294 and 1·1182 milligrams/coulomb.

If z = electrochemical equivalent of a substance in milligrams per coulomb

and I = current, in amperes, for time t seconds,
        mass of substance liberated = zIt milligrams.   (16.1)

ELEMENT	RELATIVE ATOMIC MASS	VALENCY	ELECTRO-CHEMICAL EQUIVALENT
Aluminium	27·0	3	
Chlorine	35·5	1	
Chromium	52·0	3 or 6	
Copper (cuprous)	63·6	1	
Copper (cupric)	63·6	2	0·3294 mg/C
Gold	197·2	3	
Hydrogen	1·008	1	
Iron	55·8	2 or 3	
Lead	207·2	2	
Nickel	58·7	2	0·304 mg/C
Oxygen	16·0	2	
Potassium	39·1	1	
Silver	107·9	1	1·1182 mg/C
Sodium	23·0	1	
Tin	118·7	2 or 4	
Zinc	65·4	2	0·338 mg/C

Example 16.3 *A steady current of 6·3 A is passed for 45 min through a solution of copper sulphate. Calculate the mass of copper deposited.*

Quantity of electricity, in coulombs
$$= 6·3 \text{ [A]} \times (45 \times 60) \text{ [s]}$$
$$= 17\,010 \text{ C}.$$

Since 1 coulomb deposits 0·3294 mg of copper,

∴ mass of copper deposited $= 0·3294 \text{ [mg/C]} \times 17\,010 \text{ [C]}$
$$= 5600 \text{ mg}$$
$$= 5·6 \text{ g}.$$

Example 16.4 *A current of 2 A is passed for 30 min between two platinum plates (or electrodes) immersed in a dilute solution of* *sulphuric acid in water. Calculate the mass of hydrogen and oxygen released.*

The effect of electrolysis in this case is to decompose water into its constituents, hydrogen and oxygen, the former being liberated at the negative plate (or cathode) and the latter at the positive plate (or anode).

Quantity of electricity $= 2\,[\text{A}] \times (30 \times 60)\,[\text{s}] = 3600\,\text{C}.$

From the data given in the table on page 249, the relative atomic mass and the valency of hydrogen are 1·008 and 1 respectively,

$\therefore$ chemical equivalent mass of hydrogen $= 1{\cdot}008/1 = 1{\cdot}008.$

Also, for silver, the electrochemical equivalent is 1·1182 mg/C and the chemical equivalent mass is 107·9; hence, by Faraday's second law,

electrochemical equivalent of hydrogen
$$= 1{\cdot}1182\,[\text{mg/C}] \times 1{\cdot}008/107{\cdot}9$$
$$= 0{\cdot}010\,45\,\text{mg/C},$$
$\therefore$ mass of hydrogen released
$$= 0{\cdot}010\,45\,[\text{mg/C}] \times 3600\,[\text{C}]$$
$$= 37{\cdot}6\,\text{mg} = 0{\cdot}0376\,\text{g}.$$

Similarly, chemical equivalent mass of oxygen
$$= 16/2 = 8$$
$\therefore$ electrochemical equivalent of oxygen
$$= 1{\cdot}1182\,[\text{mg/C}] \times 8/107{\cdot}9$$
$$= 0{\cdot}0829\,\text{mg/C}$$
and mass of oxygen released
$$= 0{\cdot}0829\,[\text{mg/C}] \times 3600\,[\text{C}]$$
$$= 298\,\text{mg} = 0{\cdot}298\,\text{g}.$$

Summary of Chapter 16

A current can show magnetic, heating and chemical effects. The magnetic effect is used to define the unit of current.

Mass of substance liberated from electrolyte, in milligrams
$$= z\,[\text{milligrams/coulomb}] \times Q\,[\text{coulombs}]$$
$$= z\,[\text{mg/C}] \times I\,[\text{amperes}] \times t\,[\text{seconds}] \qquad (16.1)$$

EXAMPLES 16

1. If 4 A h of electricity flow in an electrical circuit in 10 min, calculate (a) the quantity of electricity in coulombs and (b) the average value of the current.
2. If a deposit of 0·18 g of metal forms on a cathode in 10 min and a current of 1 A is flowing, what is the electrochemical equivalent of this metal?
   (U.E.I., G1)
3. A constant current of 3 A passes for 16 min through a silver coulometer. If 1 A deposits 1·118 mg of silver per second, what is the increase in the mass of the cathode? Calculate also the quantity of electricity (a) in coulombs, (b) in ampere hours.
4. A copper voltameter connected in series with an ammeter was used as a means of checking the calibration of the ammeter. During the test the ammeter reading was 2 A, and in 15 min the mass of copper deposited was 0·625 g. Take the electrochemical equivalent of copper as 0·33 mg/C and calculate (a) the true current, (b) the error in the ammeter reading and (c) the quantity of electricity which flowed during the test.   (U.E.I., G1)
5. Define the *coulomb*. In an experiment with a copper voltameter, the initial mass of the cathode was 40 g. A current of 5 A was then passed through the voltameter for one hour. If the electrochemical equivalent of copper is 0·33 mg/C, determine: (a) the number of coulombs of electricity which passed through the apparatus, (b) the new mass of the cathode.
   (U.L.C.I., G1)
6. Calculate the quantity of zinc used when a Leclanché cell supplies a current of 0·05 A for one hour. Assume the electrochemical equivalent of zinc to be 0·339 mg/C.   (N.C.T.E.C., G2)
7. A metal plate having a surface of 12 000 mm² is to be copper-plated. If a current of 2 A is passed for one hour, what thickness of copper is deposited? Density of copper is 8900 kg/m³ and the electrochemical equivalent of copper is 0·3294 mg/C.
8. It is required to deposit a layer of nickel, 0·2 mm thick, on a surface area of 15 000 mm². Calculate the minimum time required if the maximum permissible current is 8 A. Assume the electrochemical equivalent of nickel to be 0·304 mg/C and the density of nickel to be 8800 kg/m³. Also determine the quantity of electricity required (a) in coulombs and (b) in ampere hours.
9. A metal plate, having a total surface of 20 000 mm², is to be chromium-plated. If a current of 5 A is maintained for one hour, what thickness of chromium is deposited on the plate? Assume the electrochemical equivalent of chromium to be 0·09 mg/C and the relative density of chromium to be 6·6.
10. A constant current flowing through acidulated water liberates 1·248 litres of hydrogen, measured at standard temperature and pressure, per hour. Assuming the electrochemical equivalent of hydrogen to be 0·0104 mg/C and its density at standard temperature and pressure to be 90 g/m³, calculate the value of the current.
11. From the data given in the table in this chapter, determine the electrochemical equivalents of aluminium, gold and chlorine.

ANSWERS TO EXAMPLES 16

1. 14 400 C, 24 A.
2. 0·3 mg/C.
3. 3·22 g, 2880 C, 0·8 A h.
4. 2·104 A; 0·104 A, ammeter
   reading low; 1894 C.
5. 18 000 C, 45·94 g.
6. 61 mg.

7. 0·0222 mm.
8. 3·015 h; 86 800 C, 24·12 A h.
9. 0·012 28 mm.
10. 3 A.
11. 0·0932 mg/C, 0·681 mg/C,
    0·3675 mg/C.

CHAPTER 17

Electric circuit

17.1 Conductors and insulators

An Englishman, Stephen Gray, was the first person to transmit electricity a distance of about 300 metres. He did this in 1729, using a brass wire suspended by silk threads. His 'transmission line' possessed the two essential requisites for such purpose, namely (a) a good conductor to allow electrons to flow easily along it and (b) a good insulator to ensure that the movement of the electrons was confined to the conducting path along which they were to be transmitted.

It was explained in section 15.2 that materials such as copper, which are good conductors, possess a plentiful supply of free electrons, and that it is the drift of these electrons that constitutes an electric current. Materials, such as paper and rubber, possess very few free electrons and are therefore very good insulators. Hence, if a copper wire has a covering of, say, rubber or paper, it is said to be insulated, and practically none of the electrons travelling along the wire can 'leak' from it.

It is usual to divide materials into two categories, conductors and insulators; but it should be realized that these are only relative terms. No material is a perfect conductor and no material is a perfect insulator. In general, metals such as copper and aluminium are very good conductors, whereas non-metallic materials such as rubber and oil are good insulators. Air, in its normal atmospheric condition, is a very good insulator.

The insulating property of oil and of fibrous materials, such as paper and cotton, is greatly affected by the amount of moisture they contain. The greater the moisture content, the poorer is the insulating property.

Certain materials such as germanium and silicon are neither good conductors nor good insulators and are referred to as *semiconductors*. Germanium and silicon crystals, doped with certain impurities, are used as the constituents of rectifiers and transistors.

254	*Engineering Science in SI Units*

17.2 Heating effect of an electric current

The laws relating to the heating effect of an electric current were
discovered by James Prescott Joule whose name is perpetuated by
the unit of energy, the *joule* (section 1.9). The general principle of
the apparatus used by Joule is shown in fig. 17.1. A copper calorimeter B contains a known mass of water, and any loss of heat is
minimized by a layer of cotton wool C between B and an outer

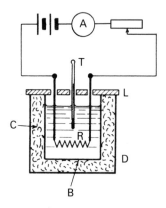

Fig. 17.1 Verification of Joule's laws.

container D. Stout copper wires pass through a wooden lid L and
are attached at the lower ends to a helix R of known resistance. The
temperature of the water is measured by a thermometer T. A
constant current is passed through R for a known time and the rise
of temperature of the water is noted. It is necessary to stir the
water very thoroughly during the test to ensure uniform distribution
of the heat generated in R.

From his researches, Joule deduced that the heat generated in a
wire is proportional to:

(*a*) the square of the current, e.g. if the current is doubled, the rate
of heat generation is increased fourfold,
(*b*) the resistance of the wire; e.g. if the length of the wire is
doubled, the rate of heat generation by a given current is also
doubled,
(*c*) the time during which the current is flowing.

Consequently, if a current I flows through a circuit having resistance R for time t,

$$\text{heat generated} \propto I^2 Rt.$$

The *unit of resistance* is that resistance in which a current of 1 ampere flowing for 1 second generates 1 joule of thermal energy. This unit of resistance is termed the *ohm* in commemoration of Georg Simon Ohm (1787–1854), the German physicist who enunciated Ohm's law (section 17.7). Hence, if a current I, in amperes, flows through a circuit having a resistance R, in ohms, for a time t, in seconds,

$$\text{heat generated, in joules} = I^2 Rt \qquad (17.1)$$

The symbol for ohm is the Greek letter Ω (capital omega). Sometimes it is more convenient to express the resistance in millionths of an ohm, in which case the resistance is said to be so many *microhms* ($\mu\Omega$). On the other hand, when we are dealing with the resistance of insulating materials, the ohm is inconveniently small; consequently, another unit called the *megohm* ($M\Omega$) is used, one megohm being a million ohms.

Example 17.1 *A current of 5 A was maintained for 6 min through a 1·3-Ω resistor immersed in 0·44 l of water. The water equivalent of the vessel and heater was 17·2 g. Assuming no loss of heat, calculate* (a) *the heat generated and* (b) *the temperature rise of the water.*

(*a*) Heat generated in resistor $= I^2 Rt$ joules
$$= 5^2 \times 1\cdot3 \times 6 \times 60 = 11\ 700\ \text{J}.$$

(*b*) Assuming the mass of 1 litre of water to be 1 kg,

mass of water in vessel $= 0\cdot44$ kg

and   water equivalent of vessel and heater
$$= 0\cdot0172\ \text{kg},$$

$\therefore$ water equivalent of water, vessel and resistor
$$= 0\cdot44 + 0\cdot0172 = 0\cdot4572\ \text{kg}.$$

From expression (13.2), heat required to raise the temperature of 0·4572 kg of water by t degrees Celsius

$$= 4190\ [\text{J/kg }^\circ\text{C}] \times 0\cdot4572\ [\text{kg}] \times t$$

But heat generated in resistor = heat absorbed by water, vessel and resistor

i.e.   11 700 [J] = 4190 [J/kg °C] × 0·4572 [kg] × t

∴   temperature rise = t = 6·11°C.

17.3 Electrical power

Since power is the rate of doing work, it follows that power is expressed in joules/second or *watts* (section 1.10); and since the electrical energy converted into thermal energy when a current I amperes flows for t seconds through a circuit having a resistance R ohms is I^2Rt joules,

∴   electrical power $= \dfrac{I^2Rt}{t}$ joules/second

$$= I^2R \text{ watts} \qquad (17.2)$$

17.4 Unit of electrical energy

In mechanics and heat we have used the SI unit of energy, namely the joule and its multiples the kilojoule and the megajoule. In electrical work, however, it has been the practice to express the energy in kilowatt hours and this practice will undoubtedly continue.

1 kilowatt hour = 1000 watt hours
= 1000 × 3600 watt seconds or joules
= 3 600 000 J = 3600 kJ = 3·6 MJ.

Example 17.2 *The wire used in a heater element has a resistance of 57 Ω. Calculate* (a) *the power, in kilowatts, when the heater is taking a current of* 3·8 *A,* (b) *the energy absorbed in* 4 *hours, in kilowatt hours,* (c) *the cost of energy consumed if the charge is* 5·5 *pence per kilowatt hour.*

(a)   Power = (3·8)² [A]² × 57 [Ω] = 823 W
=0·823 kW.

(b)   Energy absorbed = 0·823 [kW] × 4 [h]
= 3·292 kW h.

(c)   Cost of energy = 3·292 [kW h] × 5·5 [p/kW h]
                  = 18·1 p.

17.5 Fall of electrical potential along a circuit

Suppose CD in fig. 17.2 to represent a long thin wire of uniform diameter made of an alloy such as Eureka (60 per cent copper and 40 per cent nickel) having a much higher resistance than the same length and diameter of copper. The wire is connected across the

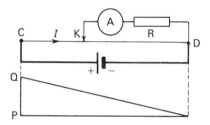

Fig. 17.2 Fall of electrical potential.

terminals of an accumulator. A milliammeter A in series with a resistor R of, say, 1000 Ω is connected between terminal D and a contact K that can be moved along CD. The function of R is to limit the current through ammeter A, thereby protecting it from an excessive current and preventing an appreciable fraction of the current being diverted from length KD of the wire.

It is found that as the distance between K and D is increased, the current through A also increases, as indicated by the height of the graph shown in fig. 17.2, where PQ represents the ammeter reading when K is at C. Since the direction of the current *I* in CD is assumed to be from C to D, C is said to be at a higher electrical *potential* than D. In other words, when two points at different electrical potentials are connected together by a conductor, electricity is assumed to flow from the one at the higher potential to that at the lower potential (section 15.4). In fig. 17.2, the difference of potential between K and D is directly proportional to the deflection on A and therefore to the distance between D and the movable contact, the latter being at a higher potential than D.

17.6 Hydraulic analogy of fall of potential

A brass tube T (fig. 17.3) has a number of glass tubes attached to it. The tube is connected to a large jar A filled with a solution, such as methylene blue, which stands out clearly against a white background. At the other end of T, there is a tap C by which the flow of the liquid can be controlled. When C is shut, the level of the liquid in the glass tubes is the same as that in the jar; i.e. the whole pressure

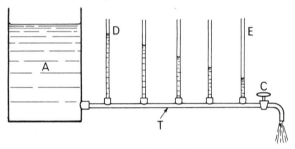

Fig. 17.3 Hydraulic analogy.

head of the liquid is available at C and there is no pressure drop in the pipe. Such a condition corresponds to a cell on open circuit, namely when there is no conducting path between the terminals.

If tap C is opened gradually, the liquid flows out at an increasing rate, and it is found that the height of the liquid in the glass tubes varies from a minimum in E to a maximum in D; in fact, if a straight rod be placed opposite the heights of the liquid columns in D and E, it will also coincide with the heights in the intermediate tubes, showing that the difference of pressure between any two points along the tube is proportional to the distance between them. Also, it is found that the more rapidly the liquid is allowed to run out at C, the greater is the difference of pressure between two adjacent tubes; in other words, the greater is the fall of pressure in a given length of pipe. These effects are somewhat similar to the electrical relationships discussed in the next section, where it is shown that the difference of potential across a resistor is proportional to the current and to the resistance.

17.7 Ohm's law

As long ago as 1827 Dr G. S. Ohm discovered that the current through a conductor, under constant conditions, was proportional to the difference of potential across the conductor. This fact can be demonstrated by connecting, as shown in fig. 17.4, a fixed resistor X, in series with a variable resistor B and an ammeter A across the terminals of an accumulator. A voltmeter* V is connected across X.

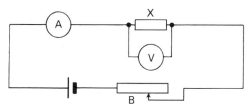

Fig. 17.4 Variation of p.d. with current.

Different currents are obtained by varying B, and for each current the reading on V is noted. It is found that the ratio

$$\frac{\text{potential difference across X}}{\text{current through X}}$$

remains constant within the limits of experimental error, i.e. the current through a circuit having a constant resistance is proportional to the difference of potential across the circuit.

The *unit of potential difference* (or p.d.) is the voltage* across a 1-ohm resistor carrying a current of 1 ampere and is termed the *volt* (V) after Count Alessandro Volta (1745–1827), an Italian physicist who was the first to discover how to make an electric battery (section 21.1). It follows from the above experiments that if the current through a resistor of 1 ohm is increased to, say, 3 A, the p.d. is 3 volts, and that if the resistance is increased to, say, 4 Ω with the current maintained at 3 A, the p.d. across the resistor becomes 3 × 4, namely 12 V. Hence if a circuit having a resistance of R ohms is carrying a current of I amperes, the p.d., V volts, across the circuit is given by:

* The commonly used moving coil, consists of a milliammeter connected in series with a resistor having a high resistance, as shown in fig. 17.2. At this stage, however, a voltmeter may be regarded merely as an instrument that indicates the difference of electric potential between the two points across which it is connected.

* The term *voltage* originally meant a difference of potential expressed in volts; but it is now used as a synonym for potential difference irrespective of the unit in which it is expressed. For instance, the voltage between the lines of a transmission system may be 400 kV, while in communication and electronic circuits, the voltage between two points may be 5 μV.

$$V = IR, \quad \text{or} \quad I = V/R, \quad \text{or} \quad R = V/I\dagger \qquad (17.3)$$

This relationship, known as *Ohm's Law*, is more complete and useful than that originally enunciated by Ohm. At this stage it is best to memorize Ohm's Law in one form only; and for this purpose, the form $I = V/R$ is probably the most convenient.

From (17.2), electrical power

$$= I^2R \text{ watts}$$
$$= I \times IR = IV \text{ watts}\ddagger \qquad (17.4)$$

or

$$= \left(\frac{V}{R}\right)^2 \times R = \frac{V^2}{R} \text{ watts} \qquad (17.5)$$

Example 17.3 *An electric kettle takes* $3\,kW$ *from a 240-V supply. Calculate the time required to raise the temperature of* $1\cdot7\,l$ *of water from* $8°C$ *to the boiling point, if the efficiency of the kettle is 85 per cent. Also calculate the value of the current and the cost of the electrical energy consumed if the tariff is 5·5 pence per kilowatt hour.*

Assuming the mass of 1 litre of water to be 1 kg,

mass of water in kettle = 1·7 kg.

Rise of temperature = 100 − 8 = 92°C.

From expression (13.2),

useful heat = 4190 [J/kg °C] × 1·7 [kg] × 92 [°C]
= 655 000 J.

Since efficiency of kettle is 85 per cent,

input energy to kettle = 655 000/0·85 = 771 000 J

$$= \frac{771\,000\,[\text{J}]}{3\,600\,000\,[\text{J/kW h}]} = 0\cdot214 \text{ kW h.}$$

† It follows from this relationship that the *ohm* can be defined as *the resistance between two points of a circuit when a constant p.d. of 1 V, applied between these points, produces in the circuit a current of 1 A, assuming that there is no source of e.m.f. in the circuit between the two points.*

‡ This relationship gives us an alternative definition of the *volt*, namely *the difference of potential between two points of a circuit carrying a constant current of 1 A when the power dissipated between these points is 1 W.*

Time required, in hours $= \dfrac{\text{energy, in kilowatt hours}}{\text{power, in kilowatts}}$

$= 0.214 \, [\text{kW h}]/3 \, [\text{kW}] = 0.0713 \, \text{h}$

$= 0.0713 \times 60 = 4.28 \, \text{min.}$

Alternatively, $771\,000 \, [\text{J}] = 3000 \, [\text{W}] \times$ time in seconds

∴   time required $= 257 \, \text{s} = 4.28 \, \text{min.}$

$\text{Current} = \dfrac{\text{power, in watts}}{\text{p.d., in volts}}$

$= 3000 \, [\text{W}]/240 \, [\text{V}] = 12.5 \, \text{A.}$

Cost of energy $= 0.214 \, [\text{kW h}] \times 5.5 \, [\text{p/kW h}]$

$= 1.2 \, \text{p.}$

Example 17.4 *An electric furnace is required to raise the temperature of 2 kg of brass from 12°C to 500°C in 15 min. The supply voltage is 230 V and the efficiency of the furnace is 80 per cent. Assuming the specific heat capacity of brass to be 370 J/kg °C, calculate (a) the energy absorbed, in kilowatt hours, (b) the power supplied to the furnace, (c) the current and (d) the resistance of the heating element.*

(*a*)   Rise of temperature $= 500 - 12 = 488°\text{C.}$

From expression (13.3),

useful energy $= 370 \, [\text{J/kg °C}] \times 2 \, [\text{kg}] \times 488 \, [°\text{C}]$

$= 361\,000 \, \text{J.}$

Since $1 \, \text{kW h} = 3\,600\,000 \, \text{J,}$

useful energy $= \dfrac{361\,000 \, [\text{J}]}{3\,600\,000 \, [\text{J/kW h}]} = 0.1002 \, \text{kW h.}$

Allowing 80 per cent for the furnace efficiency,

energy absorbed $= 0.1002/0.8 = 0.125 \, \text{kW h.}$

(*b*) Input power, in kilowatts $= \dfrac{\text{energy, in kilowatt hours}}{\text{time, in hours}}$

$= \dfrac{0.125 \, [\text{kW h}]}{(15/60) \, [\text{h}]} = 0.5 \, \text{kW.}$

(*c*) From expression (17.4),

$(0.5 \times 1000) \, [\text{W}] = I \times 230 \, [\text{V}]$

∴ $I = 2.175 \, \text{A.}$

(*d*) From expression (17.3),

$$R = 230 \text{ [V]}/2 \cdot 175 \text{ [A]}$$
$$= 105 \cdot 8 \ \Omega.$$

17.8 Electromotive force (e.m.f.)

It was shown in section 16.1 that when a current is passed for several minutes between two lead plates immersed in a dilute solution of sulphuric acid in water, a chocolate-colour coating is formed on the positive plate and that an electric current is obtained when the plates are then connected to a separate circuit; i.e. the combination of plates and acid became a voltaic cell* capable of converting chemical energy into electrical energy. Such an arrangement is a source† of an *electromotive force* (e.m.f.); in other words, an electromotive force represents something in the cell which impels electricity through a conductor connected across the terminals of that cell. Electromotive force is measured in *volts* and is represented by the symbol E. Difference of potential between two points is represented by the symbol V.

Consideration of theories accounting for the presence of an e.m.f. between the plates of a voltaic cell is outside the scope of this book. As far as we are concerned, the fact has to be accepted that when plates of different materials, such as lead and lead dioxide (as used in a lead-acid accumulator) or zinc and carbon (as used in the Leclanché primary cell), are placed in suitable solutions, an e.m.f. exists between the plates; and if a resistor is connected across them, an electric current flows through it.

Suppose E, in volts, to be the e.m.f. of cell B in fig. 17.5 and I to be the current, in amperes, when a circuit having a resistance R, in ohms, is connected across the terminals. If the *internal resistance of the cell is negligible*,‡ the terminal voltage of the cell is the same as its e.m.f., namely E; hence

$$E = IR \quad \text{or} \quad I = E/R \tag{17.6}$$

---

* See section 21.1.

† There are sources of e.m.f. other than voltaic cells, e.g., magnetic flux cutting a conductor and junctions of dissimilar metals at different temperatures.

‡ The effect of the internal resistance of a cell is considered in section 18.3.

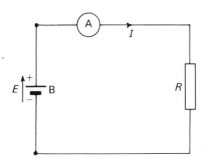

Fig. 17.5 e.m.f. of a cell.

17.9 Series connection of cells

Three accumulators were connected in series as in fig. 17.6, i.e. the negative terminal of the first cell was connected to the positive

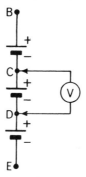

Fig. 17.6 Cells in series.

terminal of the second, and the negative of the second to the positive of the third. A voltmeter V was connected across various pairs of terminals in turn and the following results were obtained:

Terminals	Terminal voltage, volts
ED	2·16
DC	2·08
CB	2·1
EB	6·32

The sum of the terminal voltages across ED, DC and CB = 6·34 volts, which is the same—within experimental error—as the total across EB. It is therefore evident that the e.m.f. of a number of cells connected in series is the sum of their individual e.m.f.s.

17.10 Parallel connection of cells

The positive ends of two similar accumulators were joined together to one end of a resistor R as shown in fig. 17.7. The negative ends were connected through milliammeters to the other end of R. When R was adjusted to 100 ohms, the readings on A, B and C were found to be 11·3, 8·5 and 19·8 milliamperes respectively. The milliam-

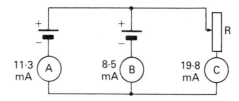

Fig. 17.7 Cells in parallel.

meters had relatively low resistance, so that the e.m.f. of the parallel cells is approximately (19·8/1000) [A] × 100 [Ω] = 1·98 V. These results indicate that when similar cells are in parallel, the e.m.f. is the same as that of one cell. On the other hand, the current is divided between the cells and the total current is the sum of the currents through the individual cells.

Summary of Chapter 17

Conductors are materials which allow electricity to pass freely, whereas insulators are materials which almost entirely obstruct the passage of electricity.

The heating effect of a current is proportional to the square of the current, to the resistance and to the time; i.e.

$$\text{heat generated} = I^2Rt \text{ joules} \qquad (17.1)$$
$$= IVt \text{ joules.}$$

$$\text{Power} = I^2 R \text{ watts} \qquad (17.2)$$
$$= IV \text{ watts} \qquad (17.4)$$
$$= V^2/R \text{ watts.} \qquad (17.5)$$

Ohm's Law:

$$I = V/R, \quad V = IR \quad \text{or} \quad R = V/I \qquad (17.3)$$

For a battery having e.m.f. E and negligible internal resistance,
$$I = E/R \qquad (17.6)$$

When two or more cells are connected in series,

$$\text{total e.m.f.} = \text{sum of their e.m.f.s.}$$

When two or more cells are connected in parallel,

$$\text{total current} = \text{sum of their currents.}$$

EXAMPLES 17

1. A certain circuit has a resistance of 50 Ω. Calculate the voltage across the circuit when the current is 3 A.
2. A certain circuit has a resistance of 800 Ω. Calculate the current when the voltage across the circuit is 2 kV.
3. When the voltage across a certain circuit is 5 mV, the current is 2 A. Calculate the resistance of the circuit in microhms.
4. Using Ohm's Law, complete the following table:

Voltage, V volts	Current, I amperes	Resistance, R ohms
25	0·25	–
–	0·75	4
240	–	120
–	0·5	1000

(U.E.I., G1)

5. A coil of insulated copper wire has a resistance of 150 Ω. If the current through the coil is 0·5 A, calculate (*a*) the p.d. across the coil and (*b*) the power absorbed by the coil.
6. Calculate the current through a lamp which is taking 100 W from a 230-V supply. Also, find the corresponding resistance of the filament.
7. How many coulombs of electricity will flow in 10 hours when a p.d. of 2 mV is applied across a resistor having a resistance of 4 μΩ?
8. If a voltmeter has a resistance of 30 kΩ, calculate the current and the power absorbed when the p.d. across the voltmeter is 460 V.
9. How many joules of energy are taken from a 12-V battery supplying 2 A for 50 min? (N.C.T.E.C., G.2)
10. An ammeter of resistance 20 Ω is connected in series with a coil whose resistance is to be measured. When a current of 3 A passes through the

coil, the potential difference across both coil and ammeter is 240 V. What is the true resistance of the resistor? What is the percentage error in calculating the resistance directly from the meter readings?

(U.L.C.I., G1)

11. How many 150-W electric lamps could safely be used on a 250-V supply which will fuse at 5 A? (U.E.I., G2)

12. How many appliances rated at 250 V and 0·5 kW could safely be run from a 13-A supply? (U.E.I., G2)

13. An electric motor takes a current of 38 A at 240 V. Neglecting all losses, calculate the output power of the motor and the number of kilojoules of work done by the motor in 10 min.

14. An electric motor connected across a 460-V supply is developing 40 kW with an efficiency of 90 per cent. Calculate (*a*) the current, (*b*) the input power and (*c*) the cost of running the motor at that load for 6 h, if the charge for electrical energy is 5 p/kW h.

15. A generator is supplying 80 lamps, each taking 60 W at 240 V. Calculate (*a*) the total current supplied by the generator, (*b*) the number of kilowatt hours consumed in 4 h, (*c*) the output power of the engine driving the generator, if the efficiency of the latter is 85 per cent.

16. The energy absorbed in 10 min by an electric heater is 1·5 MJ. The supply voltage is 240 V. Calculate (*a*) the current, (*b*) the quantity of electricity, in coulombs, taken in 5 min and (*c*) the energy, in kilowatt hours, absorbed in 100 h.

17. The heating element of an electric kettle has a resistance of 80 Ω. Calculate the time required by a current of 3·1 A to raise the temperature of 1·2 litres of water from 14°C to the boiling point, if the efficiency of the kettle is 78 per cent. Also, calculate (*a*) the power and (*b*) the cost of the energy consumed at 6·0 p/kW h.

18. A current of 2·6 A was passed through a coil of wire immersed in 0·726 kg of water for 10 min. The initial and final temperatures were 13·2°C and 17·8°C respectively. Assuming the water equivalent of the containing vessel and heater to be 24 g and neglecting any loss of heat, calculate (*a*) the resistance of the coil, (*b*) the electrical power.

19. An electric kettle takes 3 kW when the terminal voltage is 240 V. Calculate (*a*) the resistance of the heating element, (*b*) the quantity of water, in litres, which could be heated from 5°C to the boiling point in 5 min, assuming the efficiency of the kettle to be 80 per cent.

20. An electric furnace is required to raise the temperature of 3 kg of iron from 16°C to 750 °C in 20 min. The supply voltage is 240 V. The efficiency of the furnace is 76 per cent and the specific heat capacity of the iron is 500 J/kg °C. Calculate (*a*) the current, (*b*) the resistance of the heating element, (*c*) the power and (*d*) the energy absorbed in kilowatt hours.

ANSWERS TO EXAMPLES 17

1. 150 V.
2. 2·5 A.
3. 2500 μΩ.

12. 6 appliances.
13. 9·12 kW, 5472 kJ.
14. 96·6 A, 44·4 kW, 1335 p.

4. 100 Ω, 3 V, 2 A, 500 V.
5. 75 V, 37·5 W.
6. 0·435 A, 529 Ω.
7. 18 000 000 C.
8. 15·33 mA, 7·05 W.
9. 72 000 J.
10. 60 Ω, 33·3 per cent.
11. 8 lamps.

15. 20 A, 19·2 kW h, 5·65 kW.
16. 10·42 A, 3126 C, 250 kW h.
17. 723 s, 769 W, 0·925 p.
18. 3·56 Ω, 24·1 W.
19. 19·2 Ω, 1·81 litres.
20. 5·03 A, 47·7 Ω, 1·207 kW,
    0·402 kW h.

268

CHAPTER 18

Electrical resistance

18.1 Resistors in series

Two resistors, R_1 and R_2, were connected in series across a battery, as shown in fig. 18.1, the current being indicated by an ammeter A. The current is the same through two components in series. Also, voltmeters V_1, V_2 and V_3 were connected to measure the voltages across R_1, R_2 and the whole circuit respectively. The following readings were obtained:

A	V_1	V_2	V_3
0·92 ampere	2·2 volts	3·6 volts	5·8 volts

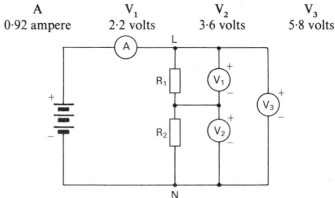

Fig. 18.1 Resistors in series.

By applying Ohm's law, we find that
$$R_1 = 2\cdot2/0\cdot92 = 2\cdot39 \ \Omega,$$
$$R_2 = 3\cdot6/0\cdot92 = 3\cdot91 \ \Omega$$
and the resistance of the whole circuit
$$= 5\cdot8/0\cdot92 = 6\cdot30 \ \Omega.$$
The sum of R_1 and R_2
$$= 2\cdot39 + 3\cdot91 = 6\cdot30 \ \Omega.$$

Hence it is seen that the total resistance of a circuit is the sum of the resistances in series; in other words, if resistors R_1, R_2 and R_3 be in series, the total resistance R is given by

$$R = R_1 + R_2 + R_3 \tag{18.1}$$

Also the voltmeters V_1, V_2 and V_3 indicate that L is at a higher potential than N by an amount equal to the sum of the potential differences across LM and MN, i.e.

$$V_3 = V_1 + V_2$$

18.2 Resistors in parallel

Two resistors R_1 and R_2 were connected in parallel as in fig. 18.2, the currents being measured by ammeters A_1 and A_2. The total current was read on A_3 and the voltage across the resistors was read on voltmeter V. It was found that the instrument readings were:

Voltmeter	A_1	A_2	A_3
5·9 volts	1·5 amperes	0·9 ampere	2·4 amperes

It will be seen that the total current is equal to the sum of the currents in the parallel circuits.

From Ohm's law, it follows that
$$R_1 = 5·9/1·5 = 3·93 \ \Omega$$
and
$$R_2 = 5·9/0·9 = 6·55 \ \Omega.$$

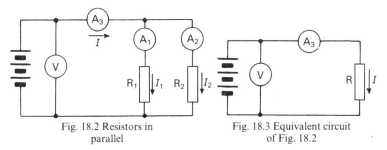

Fig. 18.2 Resistors in parallel   Fig. 18.3 Equivalent circuit of Fig. 18.2

The two resistors R_1 and R_2 in fig. 18.2 can be replaced by a single resistor R, as in fig. 18.3, the only condition being that the resistance of R must be such that the total current remains unaltered; i.e. in the above case:

$$2·4 \ [\text{A}] = \frac{5·9 \ [\text{V}]}{\text{resistance of R in ohms}}$$

∴   resistance of R $= 5·9/2·4 = 2·46 \ \Omega.$

Hence, 2·46 Ω may be said to be *equivalent* to 3·93 Ω and 6·55 Ω in parallel. It is evident that the equivalent resistance is less than either of the parallel resistances, but there does not seem to be any obvious connection between the values. For this problem it is more satisfactory to derive the relationship by considering the general case than by taking particular values.

Suppose I_1 and I_2 amperes to be the currents in parallel resistors R_1 and R_2 respectively when the p.d. is V volts (fig. 18.2). Then, by Ohm's law, $I_1 = V/R_1$ and $I_2 = V/R_2$. If I is the total current indicated by A_3,

$$I = I_1 + I_2$$
$$= \frac{V}{R_1} + \frac{V}{R_2} = V\left(\frac{1}{R_1} + \frac{1}{R_2}\right)$$

If R in fig. 18.3 represents the resistance of a single resistor through which a p.d. of V volts produces the same current I amperes, then $I = V/R$.

We have now derived two expressions for I; and by equating these expressions, we have

$$\frac{V}{R} = V\left(\frac{1}{R_1} + \frac{1}{R_2}\right)$$

$\therefore$
$$\frac{1}{R} = \frac{1}{R_1} + \frac{1}{R_2} \qquad (18.2)$$

Let us apply this expression to the experimental results considered above:

$$\frac{1}{R_1} = \frac{1}{3\cdot93} = 0\cdot254 \quad \text{and} \quad \frac{1}{R_2} = \frac{1}{6\cdot55} = 0\cdot1527$$

$\therefore$
$$\frac{1}{R} = \frac{1}{R_1} + \frac{1}{R_2} = 0\cdot254 + 0\cdot1527 = 0\cdot4067$$

and
$$R = \frac{1}{0\cdot4067} = 2\cdot46 \ \Omega,$$

which is the same as the value previously derived.

The reciprocal of the resistance, that is, 1/resistance, is termed the *conductance*, the unit of conductance being 1 *siemens* (symbol, S). From expression (18.2) it follows that for resistors connected in parallel, the conductance of the equivalent resistor is the sum of the conductances of the parallel resistors.

The current in each of the parallel resistors R_1 and R_2 in fig. 18.2 can be expressed in terms of the total current thus:

$$V = I_1 R_1 = I_2 R_2 = (I - I_1)R_2$$

$\therefore$
$$(I - I_1) = \frac{I_1 R_1}{R_2}$$

so that $\qquad I = I_1 \left(1 + \dfrac{R_1}{R_2}\right)$

$\therefore \qquad\qquad I_1 = I \cdot \dfrac{R_2}{R_1 + R_2}$ (18.3)

Similarly, $\qquad I_2 = I \cdot \dfrac{R_1}{R_1 + R_2}$

Example 18.1 *Three coils,* A, B *and* C *have resistances of* $8\,\Omega$, $12\,\Omega$ *and* $15\,\Omega$ *respectively. Find the equivalent resistance when they are connected* (a) *in series,* (b) *in parallel.*

(*a*) With the three coils in series,

$$\text{total resistance} = 8 + 12 + 15 = 35\ \Omega.$$

(*b*) If R be the equivalent resistance of the three coils in parallel, then

$$\frac{1}{R} = \frac{1}{8} + \frac{1}{12} + \frac{1}{15} = 0.125 + 0.0833 + 0.0667$$

$$= 0.275 \text{ siemens,}$$

$\therefore \qquad\qquad R = 3.64\ \Omega.$

Example 18.2 *If coils* B *and* C *of Example* 18.1 *are connected in parallel and coil* A *is connected in series, as in fig.* 18.4(a), *across a* 20-V *supply, find* (a) *the resistance of the combined circuit,* (b) *the current in each coil.*

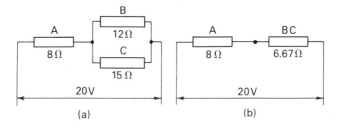

Fig. 18.4 Circuit of Example 18.2.

(*a*) Let R be the equivalent resistance of B and C, then

$$\frac{1}{R} = \frac{1}{12} + \frac{1}{15} = 0.0833 + 0.0667$$

∴ $R = 6.67\ \Omega$,

and   total resistance $= 8 + 6.67 = 14.67\ \Omega$.

(*b*) Total current $= \dfrac{20\ [V]}{14.67\ [\Omega]} = 1.364$ A, which is the current in coil A.

The p.d. across B and C in fig. 18.4(a) is the same as that across the equivalent 6.67-Ω resistor in fig. 18.4(b),

∴   total current $= \dfrac{\text{p.d. across B and C}}{\text{equivalent resistance of B and C}}$

i.e. $1.364\ [A] = \dfrac{\text{p.d. across B and C}}{6.67\ [\Omega]}$

∴   p.d. across B and C $= 1.364 \times 6.67 = 9.098$ V.

Hence,   current in B $= \dfrac{\text{p.d. across B}}{\text{resistance of B}}$

                    $= 9.098\ [V]/12\ [\Omega] = 0.758$ A

and   current in C $= 1.364 - 0.758 = 0.606$ A.

Alternatively, using expression (18.3), we have:

          current in B $= 1.364 \times 15/(12 + 15)$

                    $= 0.758$ A.

Example 18.3 *The resistance of the heating element of an electric iron is* 180 Ω. *It is connected across a 240-V supply by two conductors, each having a resistance of* 1.3 Ω. *Calculate:* (a) *the voltage across the heating element,* (b) *the voltage drop in the cable,* (c) *the power absorbed by the electric iron and* (d) *the power wasted in the cable.*

(*a*) The circuit is shown in fig. 18.5.

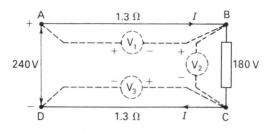

Fig. 18.5 Circuit of Example 18.3.

Total resistance of circuit = 1·3 + 180 + 1·3

= 182·6 Ω.

∴   current = $\dfrac{\text{p.d. (volts) between A and D}}{\text{resistance (ohms) between A and D}}$

= 240 [V]/182·6 [Ω] = 1·314 A.

Current also = $\dfrac{\text{p.d. between B and C}}{\text{resistance between B and C}}$

∴   1·314 [A] = $\dfrac{\text{p.d. between B and C}}{180 \ [\Omega]}$

∴   p.d. between B and C = 1·314 × 180 = 236·6 V.

(b)   Current also = $\dfrac{\text{p.d. between A and B}}{\text{resistance between A and B}}$

∴   1·314 [A] = $\dfrac{\text{p.d. between A and B}}{1·3 \ [\Omega]}$

∴   p.d. between A and B = 1·314 × 1·3 = 1·7 V.

Similarly,   p.d. between C and D = 1·314 × 1·3

= 1·7 V.

Hence,   total voltage drop in cable = 1·7 + 1·7 = 3·4 V.

The existence of these potential differences can be demonstrated by connecting moving-coil voltmeters V_1, V_2 and V_3, as shown dotted in fig. 18.5. The readings on the instruments indicate that the potential of A is 1·7 V above that of B, the potential of B is 236·6 V above that of C and the potential of C is 1·7 V above that of D. Further, it is seen that the sum of the voltmeter readings is equal to the total potential difference between A and D.

(c) Power absorbed by electric iron = 1·314 [A] × 236·6 [V]

= 311 W.

(d) Power wasted in cable = current × voltage drop in cable

= 1·314 [A] × 3·4 [V] = 4·47 W.

18.3 Effect of the internal resistance of a cell

In section 17.8 it was stated that when a cell is supplying a current and the internal resistance is negligibly small, the terminal voltage is equal to the e.m.f. of the cell. In practice, however, the resistance of the electrolyte of the cell is seldom negligible and can easily be taken into account. Thus, in fig. 18.8, TT represent the terminals

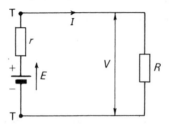

Fig. 18.6 E.M.F. of a cell

of a cell having an e.m.f. E and an internal resistance r. If a circuit of resistance R is connected across TT, then from expression (18.1), the total resistance of the circuit is $R + r$, and the current I is given by:

$$I = \frac{E}{R + r}$$

and   the terminal voltage $= V = IR$

$$= E - Ir.$$

The internal resistance of a cell can be determined by connecting a voltmeter across the terminals of the cell and noting the terminal voltage (a) with the cell on open circuit and (b) when the cell is supplying a known current I. The open-circuit reading gives the e.m.f., E, of the cell. If V be the terminal voltage when the current is I, then:

$$V = E - Ir$$

so that $r = (E - V)/I.$

Example 18.4 *Two resistors, A and B, having resistances of* $10\,\Omega$ *and* $15\,\Omega$ *respectively, are connected in parallel across a battery of four cells in series, as in fig. 16.7. Each cell has an e.m.f. of 2 V and an internal resistance of* $0.2\,\Omega$*. Calculate:* (a) *the terminal voltage of the battery,* (b) *the current through each resistor and* (c) *the total power dissipated in the resistors.*

(a)   Total e.m.f. of battery $= 2\,[V] \times 4 = 8$ V

and   total internal resistance of battery $= 0.2\,[\Omega] \times 4 = 0.8\,\Omega.$

If R is the equivalent resistance of A and B:

$$\frac{1}{R} = \frac{1}{10} + \frac{1}{15} = 0\cdot1 + 0\cdot0667$$

$$\therefore \qquad R = 6 \ \Omega.$$

Hence, total resistance of circuit $= 6 + 0\cdot8 = 6\cdot8 \ \Omega$

and
$$\text{current} = \frac{\text{total e.m.f.}}{\text{total resistance}}$$
$$= 8 \ [\text{V}]/6\cdot8 \ [\Omega] = 1\cdot177 \ \text{A}.$$

Voltage across A and B ·

    $=$ total current $\times$ equivalent resistance of A and B

    $= 1\cdot177 \ [\text{A}] \times 6 \ [\Omega] = 7\cdot06 \ \text{V}$

    $=$ potential difference across battery terminals PQ.

Alternatively, voltage across the $0\cdot8$-Ω resistor in fig. 18.7

    $= 1\cdot177 \ [\text{A}] \times 0\cdot8 \ [\Omega] = 0\cdot94 \ \text{V},$

$\therefore$ terminal voltage of battery

    $=$ battery e.m.f. $-$ voltage drop due to internal resistance

    $= 8 - 0\cdot94 = 7\cdot06 \ \text{V}.$

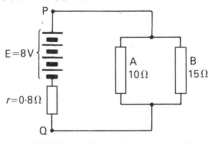

Fig. 18.7 Circuit diagram for Example 18.4.

(*b*)   Current through A $= \dfrac{\text{p.d. across A}}{\text{resistance of A}}$

    $= 7\cdot06 \ [\text{V}]/10 \ [\Omega] = 0\cdot706 \ \text{A}$

and   current through B $= 7\cdot06 \ [\text{V}]/15 \ [\Omega] = 0\cdot471 \ \text{A}$

or alternatively,

    current through B $= 1\cdot177 - 0\cdot706 = 0\cdot471 \ \text{A}.$

(*c*)   Power dissipated in resistors

    $=$ total current $\times$ p.d. across resistors

    $= 1\cdot177 \ [\text{A}] \times 7\cdot06 \ [\text{V}] = 8\cdot3 \ \text{W}.$

18.4 Relationship between the resistance and the dimensions of a conductor

For a uniform wire of a given material the resistance between any two points of the wire (i.e. the p.d. between the two points divided by the current) is proportional to the distance between the two points.

If two resistors, each of resistance R, are connected in parallel, the equivalent resistance R_e is given by:

$$\frac{1}{R_e} = \frac{1}{R} + \frac{1}{R} = \frac{2}{R}$$

$$\therefore \qquad R_e = \tfrac{1}{2}R.$$

Hence, if two wires of the same material, having the same length and diameter, are connected in parallel, the resistance of the parallel wires is half that of one wire alone. But the effect of connecting two wires in parallel is exactly similar to doubling the area of the conductor. In just the same way the effect of connecting, say, five wires in parallel is the same as increasing the cross-sectional area of a wire five times, and the result is to reduce the resistance to a fifth of that of the original wire. In general, we may therefore say that the resistance of a conductor is inversely proportional to its cross-sectional area.

Apart from the effect of temperature, referred to in section 18.5, the only other factor that influences the value of the resistance is the nature of the material; hence we may now say that:

$$\left\{ \begin{array}{c} \text{resistance} \\ \text{of a wire} \end{array} \right\} = \frac{\text{length of wire}}{\text{cross-sectional area}} \times \left\{ \begin{array}{c} \text{a constant for a} \\ \text{given material} \end{array} \right\}$$

i.e.

$$R \text{ [ohms]} = \frac{l \text{ [metres]}}{a \text{ [metres}^2]} \times \rho$$

so that

$$\rho = Ra/l \text{ ohm metres} \qquad (18.4)$$

where ρ (Greek letter, 'rho') represents the constant.

If l is 1 m and a is 1 m^2 (e.g., if the resistance is being measured between the opposite faces of a metre cube of the material), the

value of the resistance $= \dfrac{1 \text{ [m]}}{1 \text{ [m}^2]} \times \rho \text{ [}\Omega \text{ m]} = \rho$ ohm. Consequently

the constant may be regarded as the resistance of a specimen 1 m long, and 1 m^2 in cross-sectional area, and is termed the *resistivity* of the material. For example, the resistivity of annealed copper at 20°C is 0·000 000 017 25 ohm metre. It is generally more convenient to use microhms rather than ohms, so that the above value then becomes 0·017 25 $\mu\Omega$ m (or 17·25 $\mu\Omega$ mm).

Example 18.5 *Calculate the length of copper wire, 1·5 mm diameter, to have a resistance of 0·3 Ω, given that the resistivity of copper is 0·017 $\mu\Omega$ m.*

Cross-sectional area of wire $= (\pi/4) \times (1\cdot5)^2 = 1\cdot766$ mm^2
$$= 1\cdot766 \times 10^{-6} \text{ m}^2.$$

From expression (18.4), we have:

$$0\cdot3 \, [\Omega] = 0\cdot017 \times 10^{-6} \, [\Omega \text{ m}] \times \frac{\text{length}}{1\cdot766 \times 10^{-6} \, [\text{m}^2]}$$

$\therefore$ length $= 31\cdot2$ m.

18.5 Effect of temperature on resistance

Let us connect a tungsten-filament lamp L in series with an ammeter A and a variable resistor R across a 240-volt supply, as in fig. 18.8 A voltmeter V is connected across the lamp. Assuming the current through the voltmeter to be very small compared with that through the lamp, we can determine the resistance of the lamp by dividing

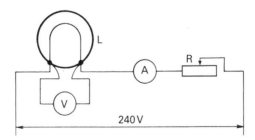

Fig. 18.8 Measurement of filament resistance at different voltages.

the voltmeter reading by the corresponding ammeter reading. By varying the value of R, we can vary the current through the lamp and thus vary the filament temperature. In this way we can determine the resistance of the tungsten filament over a wide range of temperature.

The graph in fig. 18.9 shows the results obtained with a 100-W, 240-V gas-filled lamp having a tungsten filament. It will be seen that as the temperature of the filament increases, so also does its resistance, and that the resistance at normal working temperature, i.e. with a terminal voltage of 240 V, is about ten times that of the filament when cold.

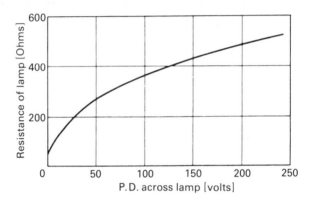

Fig. 18.9 Variation of filament resistance with voltage.

The resistance of all pure metals, such as copper, iron, tungsten, etc., increases with increase of temperature. On the other hand, the resistance of carbon, electrolytes and insulating materials, such as rubber, paper, etc., decreases with increase of temperature. The resistance of certain alloys, such as manganin (copper, manganese and nickel), remains practically constant for a considerable variation of temperature; consequently these alloys are employed whenever the resistance is required to remain practically independent of temperature, for instance in the construction of resistance boxes used for electrical measurements.

18.6 Temperature coefficient of resistance

If the resistance of a coil of insulated copper wire is measured at various temperatures up to, say, 200°C, it is found to vary as shown in fig. 18.10, the resistance at 0°C being, for convenience, taken as 1·ohm. The resistance increases uniformly with increase of temperature until it reaches 1·426 Ω at 100°C; i.e. the increase of resistance is 0·426 Ω for an increase of 100°C in the temperature, or 0·004 26 Ω/°C rise of temperature.

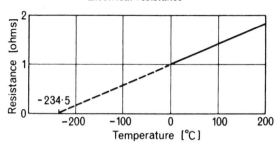

Fig. 18.10 Variation of resistance of copper with temperature.

The ratio of the increase of resistance per degree Celsius rise of temperature to the resistance at some definite temperature, adopted as standard, is termed *the temperature coefficient of resistance* and is represented by the Greek letter α. From the above figures, it follows that *if the standard temperature is assumed to be* 0°C, the temperature coefficient of resistance of annealed copper

$$= \frac{0\text{·}004\ 26\ \Omega/°\text{C}}{1\ \Omega} = 0\text{·}004\ 26/°\text{C}.$$

If the straight line of fig. 18.10 is extended backwards, the point of intersection with the horizontal axis is found to be $-234\text{·}5$°C. This means that for the range of temperature over which copper conductors are usually operated, the resistance varies as if it would be zero at $-234\text{·}5$°C. Actually, the resistance/temperature relationship is not linear below about -50°C.

In general, if a material has a resistance R_0 at 0°C and a temperature coefficient of resistance α_0 at 0°C, the increase of resistance for 1°C rise of temperature is $R_0\alpha_0$. If the temperature rises to t, the increase of resistance is $R_0\alpha_0 t$. Hence, if R_t be the resistance at t,

$$\begin{aligned} R_t &= \text{resistance at 0°C} + \text{increase of resistance} \\ &= R_0 + R_0\alpha_0 t = R_0(1 + \alpha_0 t) \qquad (18.5) \\ &= R_0(1 + 0\text{·}004\ 26\ t) \text{ for annealed copper.} \end{aligned}$$

It is usually inconvenient and unnecessary to measure the resistance at 0°C; for instance, in the case of the windings of electrical machines, it is often the practice to calculate the temperature rise after, say, three hours' operation at full load by measuring the resistance of the field coils before the commencement of the test and again immediately it is concluded. If t_1 be the initial temperature— usually taken as the temperature of the surrounding atmosphere—

and t_2 be the average temperature throughout the coils at the conclusion of the test, and if R_1 and R_2 be the corresponding resistances (fig. 18.11), then:

$$R_1 = R_0(1 + \alpha_0 t_1)$$

and

$$R_2 = R_0(1 + \alpha_0 t_2)$$

$$\therefore \quad \frac{R_2}{R_1} = \frac{1 + \alpha_0 t_2}{1 + \alpha_0 t_1} \qquad (18.6)$$

Hence, if R_1, R_2 and t_1 are known, the value of t_2 can be calculated.

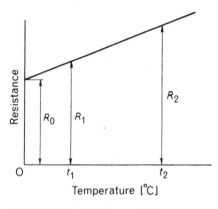

Fig. 18.11 Variation of resistance with temperature.

For rough estimate of the increase of resistance for a given increase of temperature or of the increase of temperature for a given increase of resistance, it is often the practice to assume the standard temperature to be 20°C, i.e. to assume the temperature of the atmosphere surrounding the windings to be 20°C. This means using a different value for the temperature coefficient of resistance; for instance, the temperature coefficient of resistance of annealed copper at 20°C is 0·003 92/°C. Hence for a coil of copper wire having a resistance R_{20} at 20°C, the resistance R_t at temperature t is given by:

$$R_t = R_{20}\{1 + 0·003\ 92(t - 20)\}$$

In general,

$$R_t = R_{20}\{1 + \alpha_{20}(t - 20)\} \qquad (18.7)$$

where α_{20} = temperature coefficient of resistance at 20°C.

Example 18.6 *The resistance of a coil of copper wire at the beginning of a heat test is* 173 Ω, *the temperature being* 16°C. *At the end of the test, the resistance is* 212 Ω. *Calculate the temperature rise of the coil. Assume the temperature coefficient of resistance of copper to be* 0·004 26/°C *at* 0°C.

Substituting in expression (18.6), we have:

$$\frac{212}{173} = \frac{1 + (0 \cdot 004 \ 26 \times t_2)}{1 + (0 \cdot 004 \ 26 \times 16)}$$

$\therefore$ $\qquad t_2 = 72 \cdot 5°C,$

so that temperature rise of coil = 72·5 — 16 = 56·5°C.

Example 18.7 *A certain length of aluminium wire has a resistance of* 28·3 Ω *at* 20°C. *What is its resistance at* 60°C? *The temperature coefficient of resistance of aluminium is* 0·004 03/°C *at* 20°C.

Substituting in expression (18.7), we have:

$$R_{60} = 28 \cdot 3\{1 + 0 \cdot 004 \ 03(60 - 20)\}$$
$$= 32 \cdot 86 \ \Omega.$$

Summary of Chapter 18

For resistors in series,

$$R = R_1 + R_2 + R_3 + \text{etc.} \qquad (18.1)$$

For resistors in parallel,

$$\frac{1}{R} = \frac{1}{R_1} + \frac{1}{R_2} + \frac{1}{R_3} + \text{etc.} \qquad (18.2)$$

For two resistors in parallel,

$$I_1 = I. \frac{R_2}{R_1 + R_2} \qquad (18.3)$$

$$R = \frac{\rho l}{a} \qquad (18.4)$$

The resistance of pure metals increases and that of carbon and insulating materials decreases with increase of temperature. The resistance of certain alloys, such as Eureka, is practically independent of temperature variation.

$$R_t = R_0(1 + \alpha_0 t) \tag{18.5}$$

or
$$R_t = R_{20}\{1 + \alpha_{20}(t - 20)\} \tag{18.7}$$

EXAMPLES 18

1. Two resistors, A and B, have resistances of 40 Ω and 70 Ω respectively. Calculate the resistance of the combination when they are connected (*a*) in series, (*b*) in parallel.

2. A resistor A has a resistance of 12 Ω. Calculate the resistance of another resistor B so that when A and B are in parallel, their combined resistance is 8 Ω.

3. State the THREE main effects of an electrical current, giving one practical example of each.

   Three coils have resistances of 5 Ω, 4 Ω and 2 Ω respectively.

   (*a*) Calculate the resulting resistance when they are connected in parallel.
   (*b*) If a potential difference of 12 V is applied to the circuit, what is the current taken from the supply? (N.C.T.E.C., G1)

4. Three resistors of 10 Ω, 15 Ω and 30 Ω can be connected either (*a*) in series, or (*b*) in parallel across a 140-V supply. Determine the current taken from the supply in each case and the current taken by the 30-Ω resistor in the case of the parallel connection. (U.L.C.I., G1)

5. State Ohm's law.

   Three cells each have an e.m.f. of 2 V and an internal resistance of 0·6 Ω and are connected in series, supplying current to a resistor of 8·2 Ω. Find (*a*) the current flowing, (*b*) the volt drop across the resistor.

   If the cells are now connected in parallel, find the reduction in the current flow. (E.M.E.U., G2)

6. Three resistors are to be connected in parallel to provide an equivalent network resistance of 1·5 Ω. One of the resistors is 3 Ω and another is 5 Ω. Calculate the value of the third resistor. (N.C.T.E.C., G2)

7. Two circuits, A and B, are connected in parallel across a 50-V supply. Circuit A is found to absorb 120 W, and the total current is 4·2 A. Calculate the resistances of A and B and the power absorbed by B.

8. Two resistors, 80 Ω and 20 Ω, are connected in parallel. This parallel group is connected in series with a 5-Ω resistor to a 100-V supply. Calculate the current in each resistor and the power loss in the 5-Ω resistor. (E.M.E.U., G2)

9. Three resistors are in series across a supply of 235 V and a current of 5A is flowing. If two of the resistors are 10 Ω and 25 Ω, calculate (*a*) the resistance of the third resistor and (*b*) the voltage drop across the 25-Ω resistor.

   If the 10-Ω and the 25-Ω resistors were connected in parallel, what fourth resistor in series with the third would be required to maintain the same total resistance? (U.L.C.I., G1)

10. Two resistors of 30 Ω and 20 Ω are connected in parallel. Two further resistors of 75 Ω and 50 Ω are also connected in parallel. The two groups of parallel resistors are then connected in series with an electric supply.

Make a sketch of the circuit, and if the total current in the circuit is 2 A, determine (a) the total potential difference across the circuit and (b) the current through the 50-Ω resistor. (U.L.C.I., G1)

11. A voltmeter connected across an accumulator reads 2·06 V when the cell is on open circuit. The voltmeter reading immediately falls to 1·92 V when a 0·2-Ω resistor is connected across the terminals of the cell. Calculate (a) the current, (b) the internal resistance of the cell, (c) the total electrical power generated in the cell, (d) the power dissipated in the external resistor and (e) the power wasted due to the internal resistance of the cell.

12. A battery of 10 primary cells has an e.m.f. of 15 V. When a 30-Ω resistor is connected across the battery, the terminal voltage is 12 V. Calculate (a) the current, (b) the internal resistance per cell and (c) the power dissipated in the external circuit.

13. A battery consists of 6 similar cells in series. Each cell has an e.m.f. of 2 V and an internal resistance of 0·1 Ω. The battery is connected to a lamp of resistance 3·4 Ω. Calculate the current supplied by the battery and the p.d. across the battery terminals. (U.E.I., G1)

14. Four cells, each having an e.m.f. of 1·5 V, are connected in series. Each cell has an internal resistance of 0·5 Ω. A resistor is connected across the four cells and the current in the circuit is 0·75 A. What is the terminal voltage across the four cells and what is the resistance of the resistor? (U.L.C.I., G1)

15. A circuit consists of two resistors of 6 Ω and 4 Ω respectively in parallel, connected in series with a 10-Ω resistor. The circuit is connected across a battery having an e.m.f. of 12 V and an internal resistance of 1 Ω. Determine (a) the current in the 4-Ω resistor and (b) the p.d. across the 10-Ω resistor.

Sketch a diagram of the circuit and show on it the positions in which an ammeter and a voltmeter would be placed to check the above calculated values. (U.E.I., G2)

16. The resistance of 20 m of wire of 0·1 mm diameter is 1000 Ω. Calculate the resistivity of the wire.

17. Calculate the length of wire required to make a resistor which takes 5 A from a 230-V supply. The wire is 0·1 mm in diameter and has a resistivity of 0·4 μΩ m.

18. Calculate the resistance of 1 km of aluminium wire having a diameter of 4 mm, given that the resistivity of aluminium is 0·028 μΩ m.

19. Calculate the cross-sectional area of a copper conductor, 300 m long, such that it may carry 500 A with a voltage drop of 8 V. Assume the resistivity of copper to be 0·019 μΩ m.

20. A coil consists of 400 turns of insulated copper wire having a cross-sectional area of 0·5 mm². The mean length per turn is 450 mm and the resistivity of copper at working temperature is 0·02 μΩ m. Calculate (a) the resistance of the coil and (b) the power absorbed when the p.d. across the coil is 20 V.

21. A copper rod, 6 mm in diameter and 1 m long, has a resistance of 620 μΩ. If this rod were drawn out to a wire having a cross-sectional area of 0·08 mm², what would be its resistance?

22. The field coils of a direct-current motor are wound with insulated copper wire and have a resistance of 85 Ω at 10°C. What is their resistance at 80°C? Assume the temperature coefficient of resistance of copper to be 0·0043/°C at 0°C.

23. What is meant by the *temperature coefficient of resistance* of a conductor? During a test on a conductor the following readings were obtained:

Temperature, °C	10	20	30	45	60
Resistance, Ω	3·65	3·8	3·94	4·16	4·39

Plot a graph showing the relationship between the resistance of the conductor and the temperature. From the graph, determine the temperature coefficient of resistance at 0°C.   (U.E.I., G2)

24. A constant voltage of 240 V is applied to the field windings of a motor. At 15°C the current through the windings is 2 0 A. After the motor has been running for some time, the field current was found to be 1·9 A. Calculate the temperature of the field windings given that the temperature coefficient of resistance for the coil windings is 0·0042/°C at 0°C.

(E M E.U , G2)

25. A coil of insulated copper wire has a resistance of 210 Ω at 20°C Calculate the resistance of the coil at 70°C, assuming the temperature coefficient of resistance of copper to be 0·0039/°C at 20°C.

26. An aluminium conductor is 2 km long and has a diameter of 5 mm. Calculate its resistance at 60°C, given that the resistivity of aluminium at 20°C is 0·0283 μΩ m and its temperature coefficient of resistance at 20°C is 0·004 03/°C.

ANSWERS TO EXAMPLES 18

1. 110 Ω, 25·45 Ω.
2. 24 Ω.
3. 1·053 Ω, 11·4 A.
4. 2·545 A; 28 A, 4·667 A.
5. 0·6 A, 4·92 V; 0·362 A
6. 7·5 Ω.
7. 20·83 Ω, 27·78 Ω, 90 W.
8. 0·952 A, 3 81 A, 4·762 A; 113·5 W
9. 12 Ω, 125 V, 27·86 Ω
10. 84 V 1 2 A.
11. 9 6 A, 0·0146 Ω, 19·78 W, 18·43 W, 1 35 W.
12. 0 4 A, 0·75 Ω, 4·8 W.

13. 3 A, 10·2 V.
14. 4 5 V, 6 Ω
15. 0·537 A, 8·95 V.
16. 0·3925 μΩ m.
17. 0·903 m.
18. 2·23 Ω
19. 356 mm².
20. 7·2 Ω, 55·5 W.
21. 77·3 Ω
22. 109 5 Ω.
23. 0 004 22/°C (approx.).
24. 28·3 C.
25. 251 Ω.
26. 3·35 Ω.

CHAPTER 19

Electromagnetism

19.1 Magnetic field

Before dealing with the magnetic effect of an electric current, it is necessary to explain what is meant by a magnetic field.

If a permanent magnet is suspended so that it is free to swing in a horizontal plane, as in fig. 19.1, it is found that it always takes up a position such that a particular end points towards the earth's North Pole. That end is, therefore, said to be the *north-seeking* end

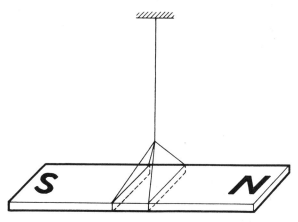

Fig. 19.1 A suspended permanent magnet.

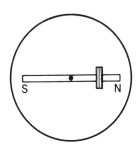

Fig. 19.2 Compass needle.

of the magnet; similarly, the other end is the *south-seeking* end. For short, these are referred to as the *north* (or N) and *south* (or S) *poles* respectively of the magnet. In the case of small compass-needles, the north pole is usually indicated by a small crosspiece, as shown in fig. 19.2, or by an arrowhead.

If the N pole of another magnet is brought near the N pole of the suspended magnet, the latter is repelled; whereas attraction occurs if the S pole of the second magnet is brought near the N pole of the suspended magnet. In general we may therefore say that like poles repel each other whereas unlike poles attract each other.

Let us next place a permanent magnet on a table, cover it over with a sheet of smooth cardboard and sprinkle some iron filings uniformly over the sheet. Slight tapping of the sheet causes the filings to set themselves in curved chains between the poles, as shown in fig. 19.3. The shape and density of these chains enable one to form a mental picture of the magnetic condition of the space or 'field' around a bar magnet and lead to the idea of *lines of magnetic flux*.

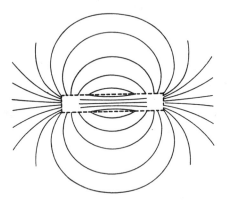

Fig. 19.3 Use of iron filings for determining distribution of magnetic field.

It is necessary to emphasize at this stage that these lines of magnetic flux have no physical existence; they are purely imaginary and were introduced by Michael Faraday as a means of visualizing the distribution and density of a magnetic field. It is important, however, to realize that the magnetic flux permeates the *whole* of the space occupied by that flux.

19.2 Direction of magnetic field

The direction of a magnetic field is taken as that in which the north-seeking pole of a magnet points when the latter is suspended in the field. Thus, if a bar magnet NS rests on a table and four compass-needles are placed in positions indicated in fig. 19.4, it is found that the needles take up positions such that their axes coincide with the corresponding chain of filings (fig. 19.3) with their north poles all

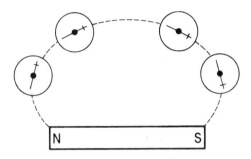

Fig. 19.4 Use of compass needles for determining direction of magnetic field.

pointing along the dotted line, from the N pole of the bar magnet to its S pole. Hence the magnetic flux is assumed to pass through the magnet, emerge from the N pole and return to the S pole. This may be expressed in another way, thus: if a compass-needle is placed near one end of a magnetized iron rod and if its N pole is repelled from the rod, the direction of the magnetic flux in that region is outwards from the rod, and the adjacent surface of the rod has a north polarity.

19.3 Characteristics of lines of magnetic flux

In spite of the fact that lines of magnetic flux have no physical existence, they do form a very convenient and useful basis for explaining various magnetic effects. For this purpose, lines of magnetic flux are assumed to have the following properties:

1. *The direction of a line of magnetic flux at any point in a non-magnetic medium, such as air, is that of the north-seeking pole of a compass-needle placed at that point,* as already described in section 19.2.

2. *Each line of magnetic flux forms a closed path*, as shown by the dotted lines in fig. 19.5 and 19.6. This means that a line of flux emerging from any point at the N-pole end of a magnet passes through the surrounding space back to the S-pole end and is then assumed to continue through the magnet to the point at which it emerged at the N-pole end.

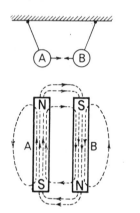

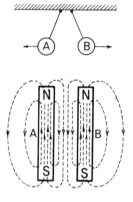

Fig. 19.5 Attraction between magnets.

Fig. 19.6 Repulsion between magnets.

3. *Lines of magnetic flux never intersect.* This follows from the fact that if a compass needle is placed in a magnetic field, its north-seeking pole will point in one direction only, namely in the direction of the magnetic flux at that point.

4. *Lines of magnetic flux are like stretched elastic cords, always trying to shorten themselves.* This effect can be demonstrated by suspending two permanent magnets, A and B, parallel to each other, with their poles arranged as in fig. 19.5. The distribution of the resultant magnetic field is indicated by the dotted lines. The lines of magnetic flux passing between A and B behave as if they were in tension, trying to shorten themselves and thereby causing the magnets to be attracted towards each other. In other words, unlike poles attract each other.

5. *Lines of magnetic flux which are parallel and in the same direction repel one another.* This effect can be demonstrated by suspending the two permanent magnets, A and B, with their N poles pointing in the same direction, as in fig. 19.6. It will be seen

that in the space between A and B the lines of flux are practically parallel and are in the same direction. These flux lines behave as if they exerted a lateral pressure on one another, thereby causing magnets A and B to repel each other. Hence like poles repel each other.

19.4 Magnetic induction and magnetic screening

In fig. 19.7, N and S are the poles of a U-shaped permanent magnet M, A and B are soft-iron rectangular blocks attached to the magnet and C is a hollow cylinder of soft-iron placed midway between A and B. The dotted lines in fig. 19.7 represent the paths of the magnetic flux due to the permanent magnet. It will be seen that this flux passes through A, B and C, making them into temporary magnets

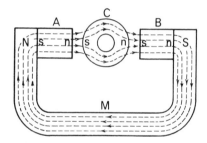

Fig. 19.7 Magnetic induction and screening.

with the polarities indicated by *n* and *s*, i.e. A, B and C are magnetized by *magnetic induction*. Being of soft-iron, A, B and C will lose almost the whole of their magnetism when they are removed from the influence of the permanent magnet M.

Fig. 19.7 also shows that no* flux passes through the air space inside cyclinder C. Consequently, a body placed in this space would be screened from the magnetic field around it. Magnetic screens are used to protect cathode-ray tubes (as used in television) and instruments such as moving-iron ammeters and voltmeters (section 19.8) from external magnetic fields.

* Actually, there must be some magnetic flux across the air space inside the soft-iron cylinder C, but the density of this flux is so low that, for most purposes, it can be assumed to be zero.

The effect of magnetic induction can also be shown by suspending a soft-iron bar A by a spring B above the poles of a U-shaped permanent magnet M, as in fig. 19.8. Bar A is magnetized by induction as shown by the dotted lines and is attracted towards M, thereby increasing the tension in B. This force of attraction may be

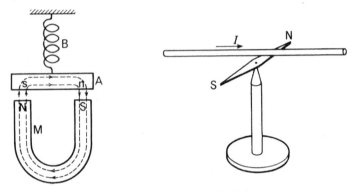

Fig. 19.8 Magnetic induction.   Fig. 19.9 Oersted's experiment.

explained as being due to the tension in the flux between A and the poles of M pulling A towards N and S, or merely as the attraction between N and *s* and between S and *n*; but these are simply two ways of stating the same thing.

19.5 Magnetic field due to an electric current

In section 16.1 it was demonstrated that one of the characteristics of an electric current is its ability to produce a magnetic effect. The discovery of this phenomenon by Oersted at Copenhagen, in 1820, was the first definite demonstration of a relationship between electricity and magnetism. Oersted found that if he placed a wire carrying an electric current above a magnetic needle (fig. 19.9) and in line with the normal direction of the latter, the needle was deflected clockwise or counterclockwise, depending upon the direction of the current. This phenomenon may be better understood from experiments made with the apparatus shown in fig. 19.10.* A stout copper wire W passes vertically through a hole in a sheet of card-

---

*It is usual to represent a current receding from the reader by a cross, as in fig. 19.10, and an approaching current by a dot, as on the left-hand conductor C in fig. 19.21. These conventions are based upon the cross being the back end

board or glass G placed horizontally, with a number of compass-needles arranged around W.

With no current through the wire, the north-seeking poles of the needles all point towards the earth's north pole. Immediately a current of, say, 10 amperes is switched on, the needles are deflected, and after a number of oscillations they come to rest with their axes lying roughly on a circle having the conductor as its centre—as shown in fig. 19.10.

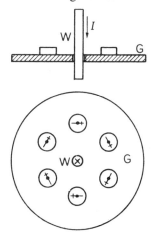

Fig. 19.10 Magnetic field of an electric current.

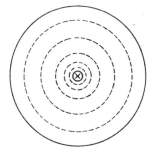

Fig. 19.11 Distribution of magnetic field in Fig. 19.10

If the current is reversed, all the compass-needles reverse their direction.

Let us now remove the compass-needles and sprinkle plate G as uniformly as possible with iron filings. A current of, say, 20 or 30 amperes is then passed through wire W and the plate tapped gently. It is found that the filings tend to arrange themselves in concentric circles around the wire (fig. 19.11), this tendency being most pronounced in the vicinity of the conductor.

These experiments indicate that a magnetic field is produced around a wire carrying an electric current and that the intensity of this field decreases as the distance from the conductor increases.

19.6 Direction of the magnetic field due to an electric current

It was pointed out in section 19.2 that the direction of the magnetic field in any particular space is always taken as the direction of the N pole of a compass-needle placed in that field. Consequently, from an experiment such as that described in connection with fig. 19.10, we can determine the relationship between the direction of the magnetic field and that of the current. Thus, it is seen from fig. 19.10 that if we look along the conductor and if the current is flowing away from us, the magnetic field has a clockwise direction.

A convenient method of representing this relationship is to grip the conductor with the *right* hand, with the thumb outstretched parallel to the conductor and pointing in the direction of the current; the fingers then point in the direction of the magnetic field around the conductor.

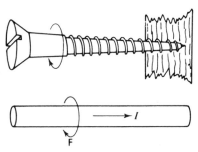

Fig. 19.12 Right-hand screw rule.

Another way of representing the relationship between the direction of a current and that of its magnetic field is to place a corkscrew or a wood-screw (fig. 19.12) alongside the conductor carrying the current. In order that the screw may travel in the same direction as the current, namely towards the right in fig. 19.12, it has to be turned clockwise when viewed from the left-hand side. Similarly, the direction of the magnetic field, viewed from the same side, is clockwise around the conductor, as indicated by curved arrow F.

19.7 Magnetic field of a solenoid

A solenoid consists of a number of turns of wire wound in the same direction, so that when the coil is carrying a current, all the turns are assisting one another in producing a magnetic field. Fig. 19.13 shows a few turns of wire wound helically through holes in a horizontal board and connected to a battery capable of supplying 15 to 20 amperes. Iron filings are evenly sprinkled over the board and the latter is gently tapped. The filings arrange themselves in concentric

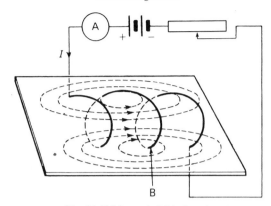

Fig. 19.13 Magnetic field of a solenoid.

paths of the shape shown by the thin dotted lines. The direction of the magnetic field can be determined by placing a compass-needle on the board and noting the direction in which its north-seeking pole is pointing. It is found that this direction is that indicated by the arrow-heads on the dotted lines in fig. 19.13.

The magnetic field may be intensified by inserting a rod of iron inside the solenoid, as shown in fig. 19.14. The magnetic flux is again represented by the dotted lines and the arrowheads indicate the direction of the flux. The iron core thus becomes magnetized with the polarities shown, and behaves like a permanent magnet so long as the current is maintained in the coil.

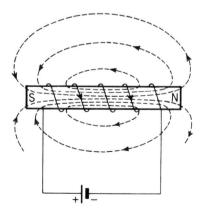

Fig. 19.14 Solenoid with an iron core.

The direction of the magnetic flux produced by a current in a solenoid may be deduced by applying either the grip or the screw rule. Thus, if the solenoid is gripped with the *right* hand, with the fingers pointing in the direction of the current, then the thumb outstretched parallel to the axis of the solenoid points in the direction of the magnetic flux *inside* the solenoid.

The screw rule can be expressed thus: If the axis of the screw is placed along that of the solenoid and if the screw is turned in the direction of the current, it travels in the direction of the magnetic flux, namely towards the right in figs. 19.13 and 19.14.

19.8 Applications of the magnetic effect of a current

The following examples are given to indicate ways in which the magnetic effect of a current may be utilized, and are not intended to be exhaustive.

(*a*) *Lifting Magnet.* A coil C, usually of insulated copper strip, is wound round a central core A forming part of an iron casting shaped as shown in fig. 19.15, where the upper half is a sectional

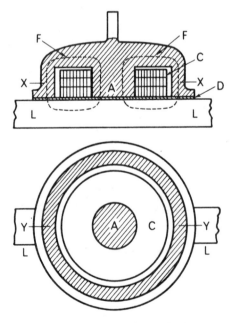

Fig. 19.15 Sectional elevation and plan of a lifting magnet.

elevation at YY and the lower half is a sectional plan at XX. Over the face of the electromagnet is a disc D of non-magnetic manganese steel which is capable of withstanding considerable impacts when the load L is picked up. The load must be of magnetic material and the dotted lines FF represent the paths of the magnetic flux.

(*b*) *Moving-iron ammeters and voltmeters.* The majority of modern moving-iron instruments, are of the *repulsion* type, in which two parallel rods or strips of soft iron, magnetized inside a coil, are regarded as repelling each other. The principal features of this type of instrument are shown in fig. 19.16, where C represents a coil wound with insulated copper wire. The number of turns on C depends upon the current required to produce full-scale deflection (see Example 19.1).

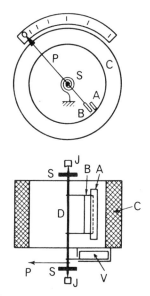

Fig. 19.16 Moving-iron instrument.

A soft-iron rod A is fixed to the bobbin on which the coil is wound and another soft-iron rod is attached to spindle D supported in jewelled bearings J. A current through coil C magnetizes rods A and B in the same direction, and B tries to move away from A because poles of the same polarity repel each other.

There are two methods of controlling the deflection of the moving system, namely:

(*a*) gravity control—now practically obsolete, and

(*b*) spring control, consisting of two spiral hairsprings S (fig. 19.16), the inner ends of which are attached to spindle D. The arrangement of these springs is shown separately in fig. 19.17. The outer end of spring B is fixed, whereas that of spring A is attached to one end of a lever L, pivoted at P, thereby enabling zero adjustment to be easily effected. The hairsprings are of non-magnetic alloy such as phosphor-bronze or beryllium-copper.

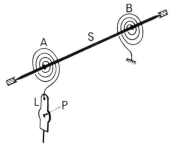

Fig. 19.17 Spring control.

With a given current through coil C (fig. 19.16), the moving system is deflected until the restoring torque exerted by the springs is equal to the deflecting torque. The value of the current is then given by the scale reading indicated by pointer P.

The tendency of the moving system to oscillate before coming to rest is damped by the thin vane V (fig. 19.16) carried by an arm attached to the spindle and moving in an air chamber.

The magnitude of the deflecting torque depends upon the product of the number of turns on coil C (fig. 19.16) and the current through the coil, i.e. it depends upon the number of *ampere turns*. For example, if full-scale deflection on a moving-iron instrument wound with 100 turns is produced by a current of 3 A,

$$\left.\begin{array}{r}\text{number of ampere turns required to}\\ \text{produce full-scale deflection}\end{array}\right\} = 3 \text{ [A]} \times 100 \text{ [t]}$$
$$= 300 \text{ A t.}$$

If a similar instrument is required to give full-scale deflection with, say, 10 A,

$$\text{number of turns} = 300 \text{ [A t]}/10 \text{ [A]} = 30.$$

It is therefore possible to arrange different instruments to have different ranges by merely winding the coils with the appropriate number of turns.

A moving-iron voltmeter is a moving-iron milliammeter connected in series with a suitable resistor, as explained in the following example.

Example 19.1 *A moving-iron instrument requires* 400 *ampere turns to give full-scale deflection. Calculate* (a) *the number of turns required if the instrument is to be used as an ammeter reading up to* 50 *A and* (b) *the number of turns and the total resistance of the instrument if it is to be arranged as a voltmeter reading up to* 300 *V with full-scale deflection when the current is* 20 *mA.*

(*a*)   Number of turns = 400 [A t]/50 [A] = 8.

(*b*)   Current to give full-scale deflection = 20 mA = 0·02 A.

∴   number of turns = 400 [A t]/0·02 [A] = 20 000.

Total resistance of voltmeter = 300 [V]/0·02 [A] = 15 000 Ω.

If the coil of 20 000 turns has a resistance of, say, 400 Ω,

resistance of the external series resistor = 15 000 − 400 = 14 600 Ω.

The advantages of moving-iron instruments are:

(i) robust construction,
(ii) relatively cheap,
(iii) can be used to measure direct and alternating currents and voltages.

The disadvantages of moving-iron instruments are:
(i) The scale divisions are not uniform, being cramped at the lower end and open at the upper end of the scale. This is due to the deflecting torque being approximately proportional to the square of the current.

(ii) Affected by stray magnetic fields. Error due to this cause is minimized by the use of a magnetic screen such as an iron casing (section 19.4).

(iii) Liable to hysteresis error when used in a direct-current circuit; i.e. for a given current, the instrument reads higher with decreasing than with increasing values of current. This error is reduced by making the iron strips of nickel-iron alloy such as Mumetal.

(iv) Moving-iron voltmeters are liable to a temperature error owing to the coil being wound with copper wire. This error is minimized by connecting in series with the coil a resistor of a material such as manganin, having a negligible temperature coefficient of resistance (section 18.6).

19.9 Force on a conductor carrying current across a magnetic field

A conductor carrying a current can produce a force on a magnet situated in its vicinity, as for example on a compass needle in fig. 19.10. From Newton's Third Law of Motion, namely that to every force there must be an opposite and equal force, it follows that a current-carrying conductor must experience a force when near a magnet. We can generalize this to—a current-carrying conductor experiences a force when in a magnetic field.

Fig. 19.18 shows a simple piece of apparatus by which this effect can be demonstrated and investigated, it being called a current balance. C is the current-carrying conductor. This is balanced on two knife edges K so that part of the conductor passes through the magnetic field NS. With the switch S open the conductor is balanced, riders R can be added (these being small pieces of wire or paper). When the switch is closed and a current flows through the conductor the balance deflects from the horizontal balance position, showing that a force is being exerted on the current-carrying conductor C.

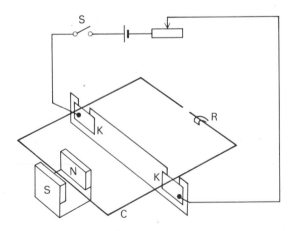

Fig. 19.18 A current balance.

The force on the current-carrying conductor can be measured by determining the weight that has to be added as rider R to restore balance. The force is found to depend on:

(i) the size of the current, the bigger the current the bigger the force;

(ii) the length of the conductor in the magnetic field, the greater the length the greater the force;

(iii) the strength of the magnetic field which is at right angles to the conductor, the stronger the field the bigger the force.

We have already in this chapter discussed the concept of magnetic flux. We can use this concept to indicate the strength of a magnetic field with the term flux density (symbol B), the greater the flux density the greater the field strength. Our current balance results can thus lead to force on conductor $\propto$ current $\times$ flux density $\times$ length of conductor

If $\quad F =$ force on conductor in newtons,

$\quad\quad I =$ current through conductor in amperes

and $\quad l =$ length, in metres, of conductor at right angles to the magnetic flux,

$\quad F$ [newtons] $\propto$ flux density $\times$ l [metres] $\times$ I [amperes].

The *unit of flux density* is taken as *the density of a magnetic flux such that a conductor carrying* 1 *ampere at right angles to that flux has a force of* 1 *newton per metre acting upon it.* This unit is termed a *tesla** (T). Hence, for a flux density B, in teslas,

$$\text{force on conductor} = BlI \text{ newtons} \quad\quad (19.1)$$

For a magnetic field having a cross-sectional area a, in square metres, and a uniform flux density B, in teslas, the *total flux*, in *webers*† (Wb), is represented by the Greek *capital* letter Φ (phi), where

$$\Phi \text{ [webers]} = B \text{ [teslas]} \times a \text{ [metres}^2\text{]}$$

or $\quad\quad\quad B \text{ [teslas]} = \dfrac{\Phi \text{ [webers]}}{a \text{ [metres}^2\text{]}} \quad\quad (19.2)$

---

* Nikola Tesla (1857–1943), a Yugoslav who emigrated to U.S.A. in 1884, was a very famous electrical inventor.

† Wilhelm Eduard Weber (1804–91), a German physicist, was the first to develop a system of absolute electrical and magnetic units.

The weber is a large unit and either the milliweber (mWb) or the microweber (μWb) is often a more convenient unit to employ, where

$$1000 \text{ milliwebers} = 1 \text{ weber}$$
and $$1\,000\,000 \text{ microwebers} = 1 \text{ weber.}$$

Example 19.2 *The pole core of an electrical machine is circular in cross-section and has a diameter of* 120 *mm. If the total flux in the core is* 16 *mWb, calculate the flux density.*

$$\text{Diameter of core} = 120 \text{ mm} = 0.12 \text{ m}$$
$$\therefore \text{ cross-sectional area of core} = (\pi/4) \times (0.12)^2$$
$$= 0.011\,32 \text{ m}^2.$$

$$\text{Total magnetic flux} = 16 \text{ mWb} = 0.016 \text{ Wb}$$
so that $$\text{flux density} = 0.016 \text{ [Wb]}/0.011\,32 \text{ [m}^2]$$
$$= 1.413 \text{ T.}$$

Example 19.3 *A conductor carries a current of* 800 *A at right angles to a magnetic field having a density of* 0.5 *tesla. Calculate the force on the conductor in newtons per metre length.*

Substituting for B, l and I in expression (19.1), we have:

$$\text{force per metre length} = 0.5 \text{ [T]} \times 1 \text{ [m]} \times 800 \text{ [A]}$$
$$= 400 \text{ N.}$$

19.10 The 'catapult' field

We can explain the force on a current-carrying conductor in terms of the pattern of magnetic flux lines. Fig. 19.19(a) shows the flux lines which would occur with a uniform magnetic field. Fig. 19.19(b) shows the flux lines that would occur with just a current-carrying conductor. Fig. 19.19(c) shows the flux pattern that occurs with a current-carrying conductor in a uniform magnetic field. We can derive this flux pattern by combining the two separate flux patterns (a) and (b). The flux lines act like stretched elastic string bent out of the straight and therefore try to return to the shortest lengths. This has the effect of exerting a downward force F on the current-carrying conductor.

We can regard the elastic string as the elastic of a catapult and in this elastic straightening out it pushes the conductor downwards.

If the current through the conductor were reversed the direction of the force would be reversed. It should be noted that the directions of the original magnetic flux, the current and the force are all mutually at right angles.

(a) Field when no current in conductor

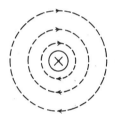

(b) Field due to only the current

(c) Field when current in conductor

Fig. 19.19

19.11 Permanent-magnet moving-coil ammeters and voltmeters

The moving-coil instrument is a good example of the application of the mechanical force exerted on a conductor carrying current across a magnetic field, discussed in the preceding section.

This type of instrument consists of a rectangular coil C (fig. 19.20) of insulated copper wire wound on a light aluminium frame which is carried by steel pivots resting in jewel bearings. Current is led into and out of the coil by spiral hairsprings AA, which also provide the controlling torque. The coil is free* to move in airgaps between the soft-iron pole-pieces PP and a central soft-iron cylinder B supported by a non-magnetic plate attached to PP.

The magnetic field is provided by a permanent magnet M of modern steel alloy such as Alcomax (iron, aluminium, cobalt, nickel and copper) to which soft-iron pole-pieces PP are attached by a special process. The hardness of modern magnet materials makes machining impossible, whereas soft iron can be easily machined to give exact airgap dimensions.

The functions of the central core B are: (a) to intensify the magnetic field by reducing the length of airgap across which the magnetic flux has to pass and (b) to give a radial magnetic flux of uniform density, thereby enabling the scale to be uniformly divided.

Damping of the moving system is effected by currents induced in the aluminium frame on which coil C is wound.

The manner in which a torque is produced when the coil is carrying a current may be understood more easily by considering a single turn PQ, as in fig. 19.21. Suppose P to carry current outwards from the paper; then Q is carrying current towards the paper. Current in

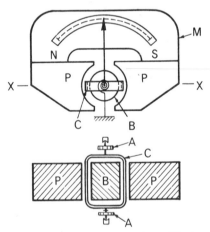

Fig. 19.20 Permanent-magnet moving-coil instrument.

* Students often form the impression that the moving coil is wound on the iron cylinder. It is important to realize that this cylinder is *fixed* and that the frame, on which the coil is wound, *does not touch* the cylindrical core.

P tends to set up a magnetic field in an anticlockwise direction around P and thus strengthens the magnetic field on the lower side and weakens it on the upper side. The current in Q, on the other hand, strengthens the field on the upper side while weakening it on the lower side. Hence, the effect is to distort the magnetic flux as shown in fig. 19.21. Since this flux behaves like a stretched elastic cord, it tries to take the shortest path between poles NS, and thus exerts forces *FF* on coil PQ, tending to move it out of the magnetic field.

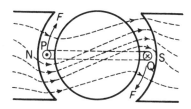

Fig. 19.21 Distribution of resultant magnetic field.

The deflecting torque ∝ current through coil

× flux density in gap

= kI for uniform flux density,

where k = a constant for a given instrument

and I = current through coil.

The controlling torque of the spiral springs ∝ angular deflection

= $c\theta$

where c = a constant for given springs

and θ = angular deflection.

For a steady deflection,

controlling torque = deflecting torque,

hence $c\theta = kI$

∴ $$\theta = \frac{k}{c}I,$$

i.e. the deflection is proportional to the current and the scale is therefore uniformly divided.

Example 19.4 *The coil of a moving-coil instrument (fig. 19.20) is wound with 42½ turns. The mean width of the coil is 25 mm and the axial length of the magnetic field is 20 mm. If the flux density in the airgap is 0·2 T, calculate the torque for a current of 15 mA.*

Since the coil has 42½ turns, one side will have 42 wires and the other side will have 43 wires.

From expression (19.1), force on the side having 42 wires

$$= 0·2 \text{ [T]} \times 0·02 \text{ [m]} \times 0·015 \text{ [A]} \times 42$$
$$= 2520 \times 10^{-6} \text{ N,}$$

∴ torque on that side of coil

$$= 2520 \times 10^{-6} \text{ [N]} \times 0·0125 \text{ [m]} = 31·5 \times 10^{-6} \text{ N m.}$$

Similarly, torque on side of coil having 43 wires

$$= 31·5 \times 10^{-6} \times 43/42 = 32·2 \times 10^{-6} \text{ N m,}$$

∴ total torque on coil

$$= (31·5 + 32·2) \times 10^{-6} = 63·7 \times 10^{-6} \text{ N m}$$
$$= 63·7 \text{ μN m.}$$

19.12 Shunts and multipliers

Owing to the delicate nature of the moving system, this type of instrument is only suitable for measuring currents up to about 50 milliamperes directly. When a larger current has to be measured, a *shunt* S (fig. 19.22), having a low resistance, is connected in parallel with the moving coil MC, and the instrument scale may be calibrated to read directly the total current I. Shunts are made of a material, such as manganin (copper, manganese and nickel), having negligible temperature coefficient of resistance. A 'swamping' resistor r, of material having negligible temperature coefficient of resistance, is connected in series with the moving coil. The latter is wound with copper wire and the function of r is to reduce the error due to the variation of resistance of the moving coil with variation of temperature.

The shunt shown in fig. 19.23 is provided with four terminals, the milliammeter being connected across the 'potential' terminals. If the instrument were connected across the 'current' terminals, there might be considerable error due to the contact resistance at these terminals being appreciable compared with the resistance of the shunt.

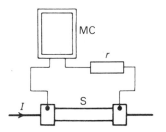

Fig. 19.22 Moving-coil instrument as an ammeter.

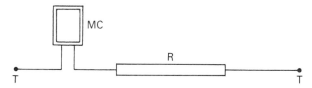

Fig. 19.23 Moving-coil instrument as a voltmeter.

The moving-coil instrument may be made into a voltmeter by connecting a resistor R of manganin or other similar material in series, as in fig. 19.23. Again the scale may be calibrated to read directly the voltage applied across terminals TT.

The main advantages of moving-coil instruments are:

  (i) high sensitivity,
  (ii) uniform scale,
  (iii) well shielded from any stray magnetic field.

The main disadvantages are:

  (i) more expensive than the moving-iron instrument,
  (ii) only suitable for direct currents and voltages.

Example 19.5 *A moving coil instrument gives full-scale deflection with 15 mA and has a resistance of 5 Ω, including that of the swamping resistor. Calculate the resistance required* (a) *in parallel to enable the instrument to read up to 1 A and* (b) *in series to enable it to read up to 10 V.*

(*a*)   Current for full-scale deflection = 15 mA = 0·015 A.
From Ohm's law,

$$\left.\begin{array}{l}\text{current through coil}\\(\text{fig. 19.22})\end{array}\right\} = \frac{\text{p.d. across coil}}{\text{resistance of coil}}$$

$$\therefore \qquad 0\cdot015 \text{ [A]} = \frac{\text{p.d. across coil}}{5 \text{ [}\Omega\text{]}}$$

so that p.d. across coil = 0·075 V.

From fig. 19.22 it follows that:

current through S = total current — current through coil
$$= 1 - 0\cdot015 = 0\cdot985 \text{ A.}$$

Similarly, current through S $= \dfrac{\text{p.d. across S}}{\text{resistance of S}}$

$$\therefore \qquad 0\cdot985 \text{ [A]} = \frac{0\cdot075 \text{ [V]}}{\text{resistance of S}}$$

so that resistance of S = 0·075/0·985 = 0·076 14 Ω.

(*b*) From Ohm's law it follows that for fig. 19.23,

current through coil $= \dfrac{\text{p.d. across TT}}{\text{resistance between TT}}$

$$\therefore \qquad 0\cdot015 \text{ [A]} = \frac{10 \text{ [V]}}{\text{resistance between TT}}$$

so that resistance between TT = 666·7 Ω.

Hence, resistance required in series with coil

= total resistance between TT — resistance of coil
$$= 666\cdot7 - 5 = 661\cdot7 \text{ }\Omega.$$

19.13 Multi-range meters

The moving-coil instrument can be arranged as a multi-range ammeter by making the shunt of different sections as shown in fig. 19.24, where A represents a milliammeter in series with a 'swamping' resistor *r* of material having negligible temperature coefficient of resistance.

With the selector switch S on, say the 50-A stud, a shunt having a very low resistance is connected across the instrument, the value of its resistance being such that full-scale deflection is produced when *I* = 50 A. With S on the 10-A stud, the resistance of the two sections of the shunt is approximately five times that of the 50-A section, and full-scale deflection is obtained when $I = 10$ A. Similarly, with S on the 1-A stud, the total resistance of the three sections is such that full-scale deflection is obtained with $I = 1$ A. Such a multi-range instrument is provided with three scales so that the value of the current can be read directly.

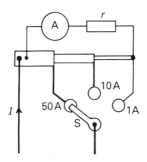

Fig. 19.24 Multi-range moving-coil ammeter.

A multi-range voltmeter is easily arranged by using a tapped resistor in series with a milliammeter A, as shown in fig. 19.25. For instance, with the data given in example 19.5, the resistance of section BC would be 661·7 Ω for the 10-V range. If D be the tapping for, say, 100 V, the total resistance between O and D = 100/0·015 = 6666·7 Ω, so that the resistance of section CD = 6666·7 — 666·7 = 6000 Ω. Similarly, if E is to be a 500-V tapping, the p.d. across section DE must be 400 V at full-scale deflection; hence the resistance of DE = 400/0·015 = 26 667 Ω. With the aid of selector switch S, the instrument can be used on three voltage ranges, and the scales can be calibrated to enable the value of the voltage to be read directly.

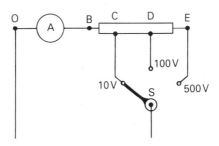

Fig. 19.25 Multi-range moving-voil
voltmeter.

19.14 The basic d.c. motor

Essentially, a d.c. motor can be considered to be just a coil in a magnetic field, as in fig. 19.26. When a current flows through the coil it experiences forces which cause it to rotate, as with the coil in the moving coil galvanometer. The coil is mounted on a shaft and so the rotation of the coil causes the shaft to rotate.

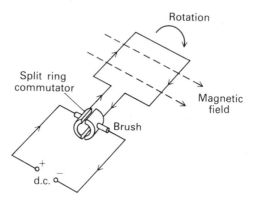

Fig. 19.26 The basic d.c. motor. For simplicity the drive shaft is not shown.

The current is fed to the coil through brushes in contact with the two parts of a split-ring commutator. As the shaft rotates the split ring rotates. This means that every time the coil, in the examples shown in the figure, passes through the vertical a brush changes its contact, from one split ring to the other and so reverses the direction of the current through the coil. This is necessary if the coil is to keep rotating in the same direction. With such a reversal that part of the coil on the left-hand side of the figure always has the current flowing in the same direction, into the paper. This means that the force acting on that side of the coil is always in the same direction.

19.15 Force between two long parallel conductors carrying electric current

It was shown in section 19.5 that a current-carrying conductor is surrounded by a magnetic field and in section 19.9 that a current-carrying conductor placed across a magnetic field has a force acting upon the conductor. It therefore follows that when two current-carrying conductors are parallel to each other, there is a force acting on each of the conductors. When the currents are in opposite directions, as in fig. 19.27(a), the two conductors repel each other. On the other hand, if the currents are in the same direction, as in fig. 19.26(b), the conductors attract each other.

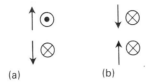

(a)   (b)

Fig. 19.27 Force between two parallel current-carrying conductors.

These effects are most easily explained by first drawing the magnetic fields produced by each conductor and then combining these fields. Thus, fig. 19.28(a) shows two conductors, A and B, each carrying current towards the paper. The magnetic flux due to the current in A alone is represented by the uniformly dotted circles in fig. 19.28(a), and that due to B alone is represented by the chain-dotted circles. It is evident that in the space between A and B the two fields tend to neutralize each other, but in the space outside A and B they assist each other. Hence the resultant distribution is somewhat as shown in fig. 19.28(b). Since magnetic flux behaves like a stretched elastic cord, the effect is to try to move conductors A and B towards each other; in other words, there is a force of attraction between A and B.

If the current in B is reversed, the magnetic fields due to A and B assist each other in the space between the conductors and the resultant distribution of the flux is shown in fig. 19.28(c). The lateral pressure in the magnetic flux exerts a force on the conductors tending to push them apart (section 19.3 (5)).

It is this force between parallel current-carrying conductors that forms the basis for the definition of the *ampere*, adopted internationally in 1948, namely, *the constant current which, if maintained in two straight parallel conductors of infinite length, of negligible circular cross-section, and placed at a distance of 1 metre apart in a vacuum, would produce between these conductors a force equal to* 2×10^{-7} *newton per metre length.*

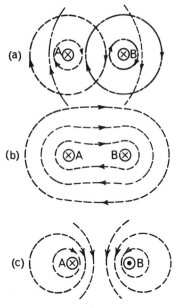

Fig. 19.28 Magnetic fields due to parallel current-carrying conductors.

Summary of Chapter 19

The characteristics of magnetic fields produced by permanent magnets and by an electric current through a straight wire and through a solenoid have been discussed, and rules are given for the direction of the magnetic field produced by an electric current.

The direction of the force exerted on a conductor carrying current across a magnetic field is deduced by deriving the distribution of the resultant flux.

Applications of electromagnetism, such as moving-iron and moving-coil instruments, are described.

For a conductor carrying current at right angles to a magnetic field,

$$\left.\begin{array}{c}\text{force on the conductor}\\\text{[newtons]}\end{array}\right\} = B\text{ [teslas]} \times l\text{ [metres]} \times I\text{ [amperes]} \qquad (19.1)$$

$$B\text{ [teslas]} = \frac{\Phi\text{ [webers]}}{a\text{ [square metres]}} \qquad (19.2)$$

EXAMPLES 19

1. Indicate, with the aid of a sketch, a method of determining the magnetic polarity of a solenoid carrying a current. (N.C.T.E.C., G2)

2. A straight conductor is situated in, and at right angles to, a uniform magnetic field. When a current is passed through the conductor there will be a force on it and the field distribution will be changed. Draw a diagram to show the direction of the current, the direction of the force and a picture of the resultant field. (U.E.I., G2)

3. Describe how you would construct an electromagnet, given a suitable soft-iron bar, insulated wire and a source of p.d. Sketch the circuit and indicate, by means of a diagram, the polarity of the magnet in relation to the direction of current flow. Explain how you would test for polarity. (E.M.E.U., G2)

4. The magnetic flux in the pole of an electric motor is 0·013 Wb. If the pole has a circular cross-section and a diameter of 120 mm, calculate the value of the flux density.

5. If the flux density inside a solenoid is 0·08 tesla and the cross-sectional area of the solenoid is 2000 mm², what is the value of the total flux in microwebers?

6. A straight conductor is carrying a current of 2500 A at right angles to a magnetic field of density 0·12 tesla. Calculate the force on the conductor in newtons per metre length.

7. A conductor, 0·3 m long, is carrying a current of 60 A at right angles to a magnetic field. The force on the conductor is 8 N. Calculate the density of the magnetic field.

8. A conductor is carrying current at right angles to a uniform magnetic field of density 1·2 teslas. The length of the conductor in the magnetic field is 150 mm. Calculate the value of the current required in order that the force on the conductor may be 20 N.

9. The coil of a moving-coil instrument is wound with $50\frac{1}{2}$ turns on a rectangular former. The axial length of the pole shoes is 23 mm and the mean width of the coil is 17 mm. If the flux density in the gap is 0·18 tesla, calculate the current required to give a torque of 30 μN m.

10. The coil of a moving-coil instrument is wound with $40\frac{1}{2}$ turns on a former having an effective length of 25 mm and an effective breadth of 20 mm. The flux density in the gap is 0·16 tesla. Calculate the torque when the current is 15 mA.

11. The armature of a certain electric motor has 900 conductors and the current per conductor is 24 A. The flux density in the airgap under the poles is 0·6 T. The armature core is 160 mm long and has a diameter of 250 mm. Assume that the core is smooth (i.e. there are no slots and the winding is on the cylindrical surface of the core) and also assume that only two-thirds of the conductors are simultaneously in the magnetic field. Calculate (*a*) the torque in newton metres and (*b*) the power developed, in kilowatts, if the speed is 700 rev/min.

   (*Note.* In the case of slotted cores, the flux density in the slots is very low, so that there is very little torque on the conductors; nearly all the torque is exerted on the teeth.)

12. A moving-coil instrument has a resistance of 5 Ω and requires a p.d. of 75 mV to give full-scale deflection.
    (a) Calculate the additional resistance to enable the instrument to read:
       (i) up to 1 A, (ii) up to 30 V.
    (b) Draw circuit diagrams showing how the resistances would be connected. (N.C.T.E.C., G2)
13. A milliammeter gives full-scale deflection with 5 mA and has a resistance of 12 Ω. Calculate the resistance necessary (a) in parallel to enable the instrument to read up to 10 A, (b) in series to enable it to read up to 100 V.
14. If the shunt for Question 13(a) is to be made of manganin strip having a resistivity of 0·5 μΩ m, a thickness of 0·5 mm and a length of 60 mm, calculate the width of the strip.
15. A moving-iron ammeter is wound with 40 turns and gives full-scale deflection with 5 A. How many turns would be required on the same bobbin to give full-scale deflection with 20 A?
    Calculate the number of turns and the total resistance of the instrument if it is arranged as a voltmeter giving full-scale reading with a current of 25 mA and a p.d. of 250 V.

ANSWERS TO EXAMPLES 19

4. 1·15 T
5. 160 μWb.
6. 300 N/m.
7. 0·444 T.
8. 111 A.
9. 8·43 mA.

10. 48·6 μN m.
11. 173 N m, 12·7 kW.
12. 0·0762 Ω, 1995 Ω.
13. 0·006 003 Ω, 19 988 Ω.
14. 9·995 mm.
15. 10 turns; 8000 turns, 10 000 Ω.

CHAPTER 20

Electromagnetic induction

20.1 Electromagnetic induction

It was mentioned in section 19.5 that the magnetic effect of an electric current was discovered by Oersted in 1820. The knowledge of this connection between electricity and magnetism caused many scientists of the time, particularly Michael Faraday in England, to try to discover a method of obtaining an electric current from a magnetic field. Failure after failure dogged Faraday's efforts until 1831 when he made the great discovery of *electromagnetic induction* with which his name will be for ever associated.

As far as we are concerned, it is more convenient to approach this matter experimentally in a different sequence from that followed by Faraday. Let us connect a coil C (fig. 20.1) to a galvanometer G, namely a very sensitive moving-coil ammeter. If a permanent

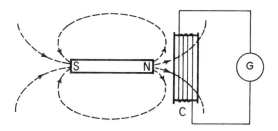

Fig. 20.1 Electromagnetic induction.

magnet NS is moved up to and along the axis of C, as shown, the moving coil of G is deflected, thereby indicating that there must be an electromotive force induced or generated in coil C. Immediately the movement of NS ceases, the moving coil of G returns to its original position. This effect proves that an e.m.f. is induced only while magnet NS is moving relative to C.

Let us now move NS away from C. The galvanometer deflection is found to be in the reverse direction, showing that the direction of the induced e.m.f. depends upon the direction in which NS is moved relative to coil C.

If, next, we hold the magnet stationary but move the coil towards the magnet and then away from it, the deflection of the galvanometer is found to follow exactly the same sequence as it did when the magnet was moved and the coil held stationary. This result shows that the generation of an e.m.f. in C depends only upon the *relative* movement of the magnet and the coil.

If the permanent magnet is turned through 180 degrees so that its S pole is pointing towards the coil, it is found that a repetition of the movements described above is accompanied by galvanometer deflections similar to those previously obtained, except that their directions are reversed. Thus, the direction of the e.m.f. induced by bringing the S pole up to the coil is the same as that previously obtained when the N pole was moved away from the coil.

The arrowheads on the dotted lines in fig. 20.1 represent the direction of the magnetic field in their respective regions. It will be seen that as the magnet is moved towards the coil, the magnetic flux of NS also moves across the wires forming the coil; that is, the magnetic flux is said to *cut* the coil. Similarly, when the coil is moved towards the magnet, the magnetic flux is said *to be cut* by the coil. It is this relative movement of the magnetic flux and the coil that causes an e.m.f. to be induced (or generated) in the latter. Alternatively,* we can say that the induced e.m.f. is due to a change in the value of the magnetic flux passing through the coil. The above experiments also show that the direction of the induced e.m.f. depends upon the direction of the magnetic flux and also upon the direction in which the coil moves relative to the magnetic flux.

Let us bring the magnet up to the coil at different speeds. It is found that the greater the speed, the greater is the deflection of the galvanometer and, therefore, the greater must be the e.m.f. induced in the coil.

---

* It is immaterial whether we consider the e.m.f. as being due to change of flux linked with a coil or due to the coil cutting or being cut by magnetic flux; the result is exactly the same. The fact of the matter is that we do not know what is really happening; but we can calculate the effect by imagining the magnetic field in the form of lines of flux, some of which expand from nothing when the field is increased or collapse to nothing when the field is reduced. In so doing, the flux may be regarded as cutting the turns of the coil, or alternatively, the effect may be regarded as being due to a change in the value of the flux passing through the coil.

20.2 Induced e.m.f.

Let us now replace magnet NS of fig. 20.1 by a coil A (fig. 20.2) connected through a switch S to a cell. At the instant when S is closed, there is a momentary deflection on G; and when S is opened, G is deflected momentarily in the reverse direction. On the other hand, if S is kept closed and coil A moved towards C, the galvanometer is deflected in the same direction as when S was closed with A

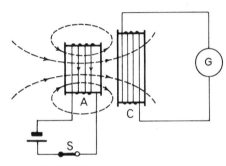

Fig. 20.2 Electromagnetic induction.

stationary. The withdrawal of A causes a deflection in the reverse direction. Deflection of G continues only while there is relative movement between the two coils, i.e. while the magnetic flux passing through coil C is changing.

The dotted lines in fig. 20.2 represent the distribution of the magnetic flux due to current in coil A. When S is opened, the current falls to zero. Consequently, the magnetic flux of A must also disappear; in other words, the magnetic flux is said to *collapse* towards A, and in so doing, the flux that passed through (or was linked with) coil C cuts the latter and induces an e.m.f. in it.

Similarly, when S is closed, the current through A causes a magnetic field to come into existence; and in this process the magnetic flux may be regarded as spreading outwards from coil A, and some of this flux will extend sufficiently to cut coil C and thereby induce an e.m.f. in it. It will be seen that as far as the e.m.f. induced in C is concerned, both the closing of S in fig. 20.2 and the moving of A towards C, with S closed, have the same effect as moving the magnet towards C in fig. 20.1.

The effects observed with the apparatus of fig. 20.2 may be accentuated by placing an iron core inside the coils, thereby increasing the magnetic flux linked with C due to a given current in A. In fact, we may go still further and wind the two coils A and C on an iron ring R, as in fig. 20.3. When S is closed, the current in A sets up magnetic flux through R, as indicated by the dotted circles. This flux, in becoming linked with coil C, induces in the latter an e.m.f. which circulates a current causing G to be deflected momentarily. So long as S remains closed, there is no further change of magnetic flux and therefore no e.m.f. induced in C. When S is opened, the magnetic flux decreases and an e.m.f. is induced in C in the reverse direction.

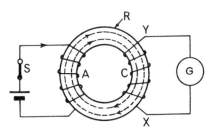

Fig. 20.3 Electromagnetic induction.

It was by means of apparatus similar to that shown in fig. 20.3 that Faraday discovered electromagnetic induction, namely that a change in the value of the magnetic flux through a coil causes an e.m.f. to be induced in that coil.

It should be pointed out that when S (fig. 20.3) is closed, the flux which becomes linked with coil C has also to grow in coil A; consequently, an e.m.f. is induced in A as well as in C. Similarly, when S is opened, the decrease of flux causes an e.m.f. to be induced in both A and C.

The results obtained from the above experiments on electromagnetic induction may now be summarized thus:

(a) When a conductor cuts or is cut by magnetic flux, an e.m.f. is induced in the conductor; or alternatively, when there is a change of magnetic flux passing through a circuit, an e.m.f. is induced in that circuit.

(b) The direction of the induced e.m.f. depends upon the direction of the magnetic flux and upon the direction in which the flux moves relative to the conductor.

(c) The magnitude of the e.m.f. is proportional to the rate at which the conductor cuts or is cut by the magnetic flux; or alternatively, the magnitude of the e.m.f. induced in a circuit is proportional to the rate of change of magnetic flux through the circuit. This last statement is often referred to as *Faraday's Law of Electromagnetic Induction*, although it was not stated in this form by Faraday.

20.3 The transformer

It is only a small step from the apparatus shown in fig. 20.3 to a transformer. The function of the latter is to change the voltage of an alternating-current supply from one value to another. An alternating voltage is maintained across coil A, termed the *primary* winding, and the alternating current through A sets up an alternating flux in the iron core. The variation of this flux causes an alternating e.m.f. to be induced in coil C (termed the *secondary* winding) as well as in coil A; and if the whole of the flux produced by the current in A passes through C, the e.m.f. induced in each turn is the same for the two coils.

Suppose the e.m.f. induced per turn to be, say, 4 V and the number of turns on the primary and secondary windings to be 50 and 500 respectively, then the e.m.f. induced in the primary winding is 200 V and that induced in the secondary winding is 2000 V. The voltage applied to the primary winding is practically equal and opposite to the e.m.f. induced in the primary and is therefore approximately 200 V. Hence such a transformer steps *up* the voltage about ten times.

Had the secondary winding been wound with only five turns, the secondary voltage would have been 20 V and the transformer would therefore step *down* the voltage to roughly a tenth of the voltage applied to the primary winding.

Since the alternating flux induces an e.m.f. in the iron core as well as in the windings, it is necessary to reduce the magnitude of the 'eddy' currents circulating in the core so as to prevent excessive loss of power in the latter. This is done by constructing the core of laminations, about 0·3 to 0·5 mm thick, insulated from one another.

20.4 Direction of the induced e.m.f.

The simplest method of determining the direction of the e.m.f. induced or generated in a conductor is to find the direction of the current due to that e.m.f. Thus, in fig. 20.4, AB represents a metal rod with its ends connected through a changeover switch S to a moving-coil galvanometer G. With S on side *a*, let us move AB *downwards* between the poles NS of an electromagnet and note the direction of G's deflection. Let us then move S over to *b*, so as to connect G in series with a high-resistance circuit R across a cell C. In order that G may again be deflected in the same direction, the polarity of C must be that shown in fig. 20.4; i.e., the current through the galvanometer must be in the direction indicated by the arrow alongside G. Hence, the e.m.f. generated in AB must be acting from A towards B when the rod is moved downwards through the magnetic field between poles NS.

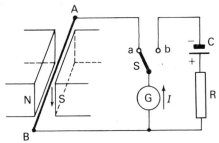

Fig. 20.4 Direction of induced e.m.f.

Now arises the problem: how can we remember this relationship in a form that can be easily applied to any other case? Two methods are available for this purpose, namely:

(a) Fleming's* right-hand rule *If the first finger of the right hand be pointed in the direction of the magnetic flux, as in fig.* 20.5, *and if the thumb be pointed in the direction of motion of the conductor* **relative** *to the magnetic field, then the second finger, held at right angles to both the thumb and the first finger, represents the direction of the e.m.f.* The manipulation of the thumb and fingers and their association with the correct quantity present some difficulty to many students. Easy manipulation can only be acquired by experience; and it may be helpful to associate *f*ield or *f*lux with *f*irst finger, *m*otion of the conductor relative to the field with the *m* in thu*m*b and *e*.m.f. with the *e* in s*e*cond finger. If any two of these are correctly applied, the third is correct automatically.

(b) Lenz's law In 1834, almost immediately after the discovery of induced currents, Heinrich Lenz, a German physicist (1804–65),

---

* John Ambrose Fleming (1849-1945) was Professor of Electrical Engineering at University College, London.

gave a simple rule, known as Lenz's law, which can be expressed thus: *The direction of an induced e.m.f. is always such that it tends to set up a current opposing the motion or the change of flux responsible for inducing that e.m.f.*

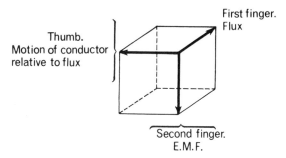

Fig. 20.5 Fleming's right-hand rule.

Let us consider the application of this law to the experiment described in connection with fig. 20.4. The current due to the e.m.f. induced in AB tends to set up an anticlockwise magnetic flux around the rod, so that the resultant flux in the vicinity of AB is distorted in the opposite direction to that shown in fig. 19.19(c). But such a distorted flux exerts an upward force upon the conductor, trying to oppose its downward movement and therefore trying to prevent that which is responsible for the generation of the e.m.f. Hence, when Lenz's law is applied to such an example, it is necessary to find the direction of the current which will distort the flux in such a direction as to try to prevent the relative movement of the conductor and the magnetic flux. The direction of such a current is also the direction of the generated e.m.f.

Let us also consider the application of Lenz's law to the ring shown in fig. 20.3. By applying either the screw or the grip rule given in section 19.6, we find that when S is closed and the cell has the polarity shown, the direction of the magnetic flux in the ring is clockwise. Consequently, the current in C must be such as to try to produce a flux in an anticlockwise direction, tending to oppose the growth of the flux due to A, namely the flux which is responsible for the e.m.f. induced in C. But an anticlockwise flux in the ring would require the current in C to be passing through the coil from X to Y (fig. 20.3). Hence, this must also be the direction of the e.m.f. induced in C.

20.5 Magnitude of the generated or induced e.m.f.

Fig. 20.6 represents the elevation and plan of a conductor AA situated in an airgap between poles NS. Suppose AA to be carrying a current I, in amperes, in direction shown. By applying either the screw or the grip rule of section 19.6, it is found that the effect of this current is to strengthen the flux on the right and weaken that on the left of A, so that there is a force of BlI newtons (section 19.9) urging the conductor towards the left, where B is the flux density in teslas (or webers per square metre) and l is the length in metres of

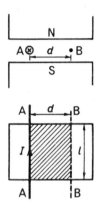

Fig. 20.6 Conductor moved across magnetic field.

the conductor in the magnetic field. Hence, a force of this magnitude has to be applied in the opposite direction to move A towards the right.

The work done in moving conductor AA through a distance d metres to position BB in fig. 20.6 is $(BlI \times d)$ joules. If this movement of AA takes place at a uniform velocity in t seconds, the e.m.f. induced in the conductor is constant at, say, E volts. Hence the electrical power generated in AA is IE watts and the electrical energy is IEt watt seconds or joules. Since the mechanical energy expended in moving the conductor horizontally across the gap is all converted into electrical energy, then

$$IEt = BlId$$

$$\therefore \quad E = \frac{Bld}{t} \text{ volts.}$$

But *Bld* = the total magnetic flux Φ, in webers, in the area shown shaded in fig. 20.6, and is therefore the flux cut by the conductor when the latter is moved from AA to BB. Hence

$$E \text{ [volts]} = \frac{\Phi \text{ [webers]}}{t \text{ [seconds]}} \tag{20.1}$$

i.e. the e.m.f., in volts, generated in a conductor is equal to the rate* (in webers/second) at which the magnetic flux is cutting or being cut by the conductor; and the *weber* may therefore be defined as *that magnetic flux which, when cut at a uniform rate by a conductor in* 1 *second, generates an e.m.f. of* 1 *volt.*

Example 20.1 *Calculate the e.m.f. generated in the axle of a car travelling at* 90 *km/h, assuming the length of the axle to be* 1·8 *m and the vertical component of the density of the earth's magnetic field to be* 40 *microteslas.*

Speed of car = 90 × 1000/3600 = 25 m/s.

Vertical component of the density of the earth's field
$$= 40 \ \mu\text{T} = 40 \ \mu\text{Wb/m}^2,$$
∴ rate at which axle cuts magnetic flux
$$= 40 \ [\mu\text{Wb/m}^2] \times 25 \ [\text{m/s}] \times 1 \cdot 8 \ [\text{m}]$$
$$= 1800 \ \mu\text{Wb/s}$$
so that e.m.f. generated in axle
$$= 1800 \ \mu\text{V}.$$

20.6 Magnitude of e.m.f. induced in a coil

Suppose the magnetic flux through a coil of N turns to be increased by Φ webers in t seconds due to, say, the relative movement of the coil and a magnet (fig. 20.1). Since the magnetic flux cuts each turn, one turn can be regarded as a conductor cut by Φ webers in t seconds; hence, from expression (20.1), the average e.m.f. induced in each turn is Φ/t volts. The current due to this e.m.f., by Lenz's law, tries to prevent the increase of flux, i.e. tends to set up an opposing flux. Thus, if the magnet NS in fig. 20.1 is moved towards coil C, the flux passing from left to right through the latter is

---

* Because it is the rate of change of flux the equation is generally written as $E = -d\Phi/dt$, the minus sign being a consequence of Lenz's law.

increased. The e.m.f. induced in the coil circulates a current in the direction represented by the dot and cross in fig. 20.7, where—for simplicity—coil C is represented as one turn. The effect of this current is to distort the magnetic field as shown by the dotted lines, thereby tending to push the coil away from the magnet. By Newton's Third Law of Motion, there must be an equal force tending to oppose the movement of the magnet towards the coil.

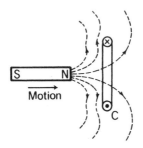

Fig. 20.7 Distortion of magnetic field caused by induced current.

Owing to the fact that the induced e.m.f. circulates a current tending to oppose the increase of flux through the coil, its direction is regarded as negative; hence

average e.m.f. induced in 1 turn $= -\Phi/t$ volts
$$= -\text{average rate of change of flux in webers per second}$$
and average e.m.f. induced in coil $= -N\Phi/t$ volts   (20.2)
$$= -\text{average rate of change of flux-linkages per second.}$$

From expression (20.2) we can define the *weber as that magnetic flux which, linking a circuit of one turn, induces in it an e.m.f. of 1 volt when the flux is reduced to zero at a uniform rate in 1 second.* This is an alternative to the definition already given in the previous section.

Example 20.2 *A magnetic flux of* 400 *μWb passing through a coil of* 1200 *turns is reversed in* 0·1 *s. Calculate the average value of the e.m.f. induced in the coil.*

The magnetic flux has to decrease from 400 μWb to zero and then increase to 400 μWb in the reverse direction;

hence the change of flux $= -800$ μWb $= -800 \times 10^{-6}$ Wb.

Substituting in expression (18.2), we have:

$$\left.\begin{array}{l}\text{average e.m.f.}\\ \text{induced in coil}\end{array}\right\} = -\frac{1200 \text{ [turns]} \times (-800 \times 10^{-6}) \text{ [Wb]}}{0{\cdot}1 \text{ [s]}}$$
$$= 9{\cdot}6 \text{ V.}$$

This e.m.f. is positive because its direction is the same as the original direction of the current, at first tending to prevent the current decreasing and then tending to prevent it increasing in the reverse direction.

20.7 An elementary a.c. generator

An elementary a.c. generator can be considered to be a coil rotating in a magnetic field (fig. 20.8). Since the flux linked by the coil is continually changing an e.m.f. is induced in the coil. For the arrangement

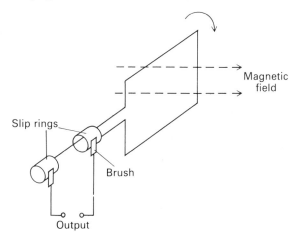

Fig. 20.8 An elementary a.c. generator.

shown in the figure the flux linked by the coil changes roughly in the form shown by the graph in fig. 20.9(a). Since the induced e.m.f. is the rate of change of linked flux then the e.m.f. is a maximum when the flux changes at its maximum rate and zero when the rate of change of flux with time is zero. The rate of change of flux with time is the slope (gradient) of the flux–time graph and so maximum e.m.f. occurs when the flux value is zero, the slope being a maximum, and zero e.m.f. when the flux value is either a maximum or a minimum, the slope then being zero. Because relative to the coil the direction of the linked flux changes then the direction of the induced e.m.f. also changes. The result is the e.m.f.–time graph shown in fig. 20.9(b).

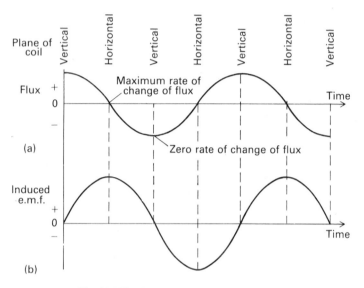

Fig. 20.9 The flux and induced e.m.f. changes.

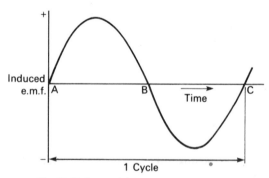

Fig. 20.10 One cycle of an alternating e.m.f.

One complete variation of the e.m.f., as shown in fig. 20.10, is termed a *cycle*, and the duration of 1 cycle is termed its *period* (or *periodic time*). The number of such cycles that occur in 1 second is referred to as the *frequency*, and a frequency of 1 cycle per second is termed 1 *hertz* (in memory of Heinrich Hertz (1857–95), a German physicist, whose laboratory experiments confirmed Maxwell's electromagnetic theory of waves).

Summary of Chapter 20

An e.m.f. is induced in a coil when a magnet or another coil carrying a current is moved either towards or away from that coil. The same effect may also be obtained by moving the first coil towards or away from the magnet or the second coil. An e.m.f. can also be induced in a coil by switching on or off the current in an adjacent coil. In general, an e.m.f. is induced in a coil whenever the latter cuts or is cut by magnetic flux, i.e. whenever there is a change of flux through the coil.

It is shown how Fleming's right-hand rule and Lenz's law can be applied for determining the direction of the induced e.m.f.

Average e.m.f. generated in a conductor $= \Phi/t$ volts   (20.1)

where Φ = flux, in webers, cut in t seconds.

Average e.m.f. induced in a coil of N turns $= N\Phi/t$ volts   (20.2)

where Φ = change of flux, in webers, in t seconds.

EXAMPLES 20

1. A conductor, 500 mm long, is moved through a magnetic field at 0·5 m/s perpendicular to the flux. What is the flux density when an e.m.f. of 0·1 V is generated?   (E.M.E.U., G2)
2. A generator conductor having a length of 200 mm at radius 100 mm is moving at 573 rev/min at right angles to the magnetic-field which has a flux density of 1·4 Wb/m². If the conductor current is 30 A, what is the e.m.f. induced, and what is the force on the conductor?   (U.E.I., G2)
3. A conductor, 120 mm long, lies at right angles to a magnetic field having a uniform density of 0·4 T. Calculate the speed at which the conductor must be moved in a direction at right angles to its length and to the magnetic field in order that an e.m.f. of 0·8 V may be generated in the conductor.
4. The axle of a certain motor car is 1·5 m long. Calculate the e.m.f. generated in the axle when the car is travelling at 140 km/h along a level road. Assume the vertical component of the density of the earth's magnetic field to be 40 μT.
5. A wire, 200 mm long, is moved at a uniform speed of 6 m/s at right angles to its length and to a magnetic field. Calculate the density of the magnetic field if the e.m.f. generated in the wire is 0·4 V.

If the wire forms part of a closed circuit having a resistance of 0·1 Ω, calculate the force on the wire.

6. A copper disc, 300 mm in diameter, is rotated at 200 rev/min about a horizontal axis through its centre and perpendicular to its plane. If the axis points magnetic north and south, calculate the e.m.f. between the circumference of the disc and the axis. Assume the horizontal component of the density of the earth's field to be 18 μT.

7. A coil of 1500 turns produces a magnetic flux of 2500 μWb when carrying a certain current. If this current is reversed in 0·2 s, what is the average value of the e.m.f. induced in the coil?

8. Two coils, A and B, are wound on the same iron core. There are 300 turns on A and 2800 turns on B. A current of 4 A through coil A produces a flux of 800 μWb in the core. If this current is reversed in 0·02 s, what are the values of the average e.m.f.s induced in A and B?

ANSWERS TO EXAMPLES 20

1. 0·4 T.

2. 1·68 V, 8·4 N.

3. 16·67 m/s.

4. 2·334 mV.

5. 0·333 T, 0·267 N.

6. 4·23 μV.

7. 37·5 V.

8. 24 V, 224 V.

CHAPTER 21

Primary and secondary cells

21.1 A simple voltaic cell

It was in 1789 that a chance observation by Luigi Galvani* (1737–98) led to the idea of generating an electric current from a source of chemical energy. Galvani, a professor of anatomy at Bologna, noticed that recently-skinned frogs' legs, hung by copper wire to an iron balcony, were convulsed whenever they touched the iron. Another Italian, Alessandro Volta* (1745–1827), a professor of physics at Pavia, subsequently showed that if a rod of two dissimilar metals, such as copper and iron, was placed so that one end was in contact with a nerve on a frog's leg and the other end in contact with a muscle on the foot, muscular contraction took place. Following his investigations of this phenomenon, Volta, in 1799, constructed a simple battery—known as Volta's pile—by assembling discs of zinc, cloth soaked with brine, and copper, piled upon one another in that order. By this means, a large number of cells were obtained in series, giving a high electromotive force, but having the disadvantage of high internal resistance.

It was a comparatively short step to replace the wet cloth of Volta's pile by an electrolyte to give the simple voltaic cell shown in fig. 21.1. In this cell, plates of copper and zinc are immersed in dilute sulphuric acid contained in a glass vessel.

When a resistor R and an ammeter A are connected in series across the terminals, it is found that current flows through R from the copper electrode to the zinc electrode and that the difference of potential between the plates is about 0·8 V, the copper plate being the positive electrode.

It is also found that when current is flowing, hydrogen gas is released in the form of bubbles on the surface of the copper plate. This gas film sets up a back electromotive force and also increases the internal resistance of the cell. These effects are known as

---

* Galvani's name is perpetuated in such terms as 'galvanize' and 'galvanometer', and the unit of e.m.f. and of potential difference, the 'volt', has been named after Volta (see section 17.7).

polarization. Many types of cells were developed to reduce polarization but they are now all obsolete except the 'dry' Leclanché cell used in electric torches, etc., and the mercury cell (sections 21.2 and 21.3).

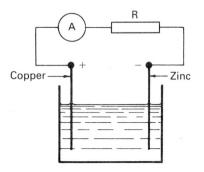

Fig. 21.1 A simple voltaic cell.

In these cells it is only possible to transform chemical energy into electrical energy, and a cell can be replenished only by renewal of the active materials. This type is usually referred to as a *primary cell.* In a *secondary cell,* the chemical action is reversible, i.e. chemical energy is converted into electrical energy when the cell is discharging, and electrical energy is converted into chemical energy when the cell is being charged, as already referred to in section 16.1.

21.2 Leclanché cell

The 'wet' type of Leclanché cell (now practically obsolete) consists of a carbon plate, surrounded by a mixture of manganese dioxide (MnO_2) and powdered carbon, in an unglazed earthenware pot. This pot and an amalgamated zinc rod are immersed in a saturated solution of salammoniac (ammonium chloride, NH_4Cl) in water. The carbon plate and the zinc rod form the positive and negative electrodes respectively.

The function of the manganese dioxide is to reduce polarization by combining with the hydrogen released at the carbon plate to form water and a brown oxide of manganese (Mn_2O_3), thus:

$$H_2 + 2MnO_2 = Mn_2O_3 + H_2O.$$

In the 'dry' type of Leclanché cell, the same ingredients are present, and fig. 21.2 is a sectional view of the construction most commonly used. A carbon rod A is surrounded by a black depolarizing paste B, consisting of manganese dioxide, powdered carbon, salammoniac, zinc chloride and water, the paste being usually contained in a bag of coarse linen. Around this depolarizer is a mixture P of flour, plaster of Paris, salammoniac and zinc chloride, with water added to form a white paste. The latter need only be thick enough to prevent the black paste B touching the zinc container Z. Above the depolarizer and the white paste is a layer S of sawdust or similar porous material, the cell being sealed with a layer of pitch T in which there is a vent tube V. The zinc container Z is usually covered with a cardboard case.

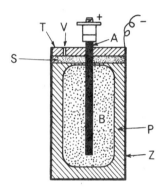

Fig. 21.2 Dry Leclanché
cell

A Leclanché cell has an e.m.f. of about 1·5 V when new, but this e.m.f. falls fairly rapidly if the cell is in continuous use. This fall is due to polarization—the hydrogen film at the carbon electrode forms faster than can be dissipated by the depolarizer. However, if the cell is disconnected from the external circuit, depolarization continues and the e.m.f. recovers its normal value. Hence the Leclanché cell is suitable only for intermittent use, e.g. for electric torches, radio receivers, etc.

The life of a dry Leclanché cell is reduced by local action and even the shelf life is limited to about two years owing to the local action that goes on continuously in the cell. Also, this type of cell does not lend itself to miniaturization, since the number of ampere hours obtainable from the cell falls off rapidly as the size is reduced.

21.3 Mercury cell*

This type of cell was developed to meet the requirements of minia-turization, e.g. for guided missiles, medical electronics, hearing aids, etc., where it is necessary to reduce the size of the cell but at the same time obtain: (*a*) a high ratio of output energy/mass; (*b*) a constant e.m.f. over a relatively long period; and (*c*) a long shelf life, i.e. absence of local action.

A cross-section of the basic type of mercury cell is shown in fig. 21.3. The negative electrode is zinc, either as a foil or as powder compressed into a hollow cylinder. This electrode is surrounded by a layer of electrolyte consisting of a concentrated aqueous solution of potassium hydroxide (KOH) and zinc oxide (ZnO). Surrounding the electrolyte is a layer of mercuric oxide (HgO). This oxide con-tains a small percentage of finely powdered graphite to reduce the internal resistance of the cell.

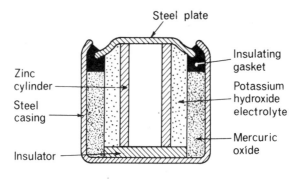

Fig. 21.3 A mercury cell.

The above constituents are assembled in a nickel-plated or stain-less-steel cylinder which forms the positive electrode. The zinc cylinder and the electrolyte are supported on a disc of insulating material; and the cell is sealed by an insulating gasket between the container and a nickel-plated steel plate resting on top of the zinc cylinder.

When the cell is supplying current, no gases are evolved at either electrode, except under abnormal operating conditions. Conse-quently there is no polarization so that the cell is able to maintain its terminal voltage practically constant at about 1·2–1·3 V (depend-

* The authors are indebted to Mallory Batteries, Ltd, for information concerning this cell.

ing upon the value of the load current) for a relatively long time. Also, owing to local action being practically negligible, the cell can be stored for a long period at normal atmospheric temperature without appreciable loss of capacity.

21.4 Secondary cells

Whenever a battery is required to supply a relatively large amount of power, secondary cells must be used. Such batteries may supply power in telephone exchanges, emergency lighting for hospitals, etc. They also form portable sources of power for starting, ignition and lighting of motor vehicles and as motive power for electrically propelled vehicles.

Secondary cells may be divided into two types: (*a*) the lead–acid cell, in which lead plates covered with compounds of lead are immersed in a dilute solution of sulphuric acid in water; (*b*) the nickel–cadmium and the nickel–iron alkaline cells.

These cells will now be considered in greater detail.

21.5 Lead–acid cell

The plates used in this type of cell may be grouped thus:

(*a*) *Formed* or *Planté* plates, namely those formed from lead plates by charging, discharging, charging in reverse direction, etc., a number of times, the forming process being accelerated by the use of suitable chemicals. The main difficulty with this type of construction is to secure as large a working surface as possible for a given mass of plate. One method of increasing the surface area is to make the plates with deep corrugations, as in fig. 21.4, with reinforcing ribs at intervals.

Fig. 21.4 Section of a
Planté plate.

(*b*) *Pasted* or *Faure* plates, namely those in which a paste of the active material is either pressed into recesses in a lead–antimony grid or held between two finely perforated lead sheets cast with ribs and flanges so that the two sheets, when riveted together, form in effect a number of boxes which hold the paste securely in position. The paste is usually sulphuric acid mixed with red lead (Pb_3O_4) for the positive plates and with litharge (PbO) for the negative plates. A small percentage of a material such as powdered pumice is added to increase the porosity of the paste. For a given ampere hour capacity, the mass of a pasted plate is only about a third of that of a formed plate.

Students are advised to examine specimen plates or plates taken from disused accumulators.

When weight is of no importance it is common practice to make the positive plates of the 'formed' type and the negative plates of the 'pasted' type. The active material on the positive plates expands when it is subjected to chemical changes; consequently it is found that the greater mechanical stiffness of the 'formed' construction is an important advantage in reducing the tendency of the plates to buckle. This tendency to buckle is reduced still further by constructing the cell with an odd number of plates, as shown in fig. 21.5, the

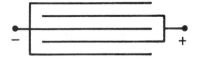

Fig. 21.5 Arrangement
of plates.

outer plates being always negative. This arrangement enables both sides of each positive plate to be actively employed, and the tendency of one side of a plate to expand and cause buckling is neutralized by a similar tendency on the other side.

The plates are assembled in glass, polystyrene, vulcanized-rubber or resin-rubber containers and separated by a special grade of paper or microporous sheets of a plastic material.

The most suitable relative density of the acid depends upon the type of cell and the state of charge of the cell. An average value, however, is about 1·21.

The terminal voltage of a lead–acid cell is about 2 V to 1·85 V when it is being discharged and about 2·1 V to 2·6 V when being charged.

21.6 Chemical reactions in a lead–acid cell

The chemical reactions taking place during charge and discharge are complicated, and all we can do here is to indicate the most important reactions and to account for the variation in the density of the electrolyte.

When the cell is fully charged, the active material on the positive plate is lead dioxide (PbO_2) and that on the negative plate is spongy or porous lead (Pb). During discharge, the lead dioxide and the spongy lead are converted into lead sulphate ($PbSO_4$). These chemical reactions are accompanied by the decomposition of some of the sulphuric acid molecules and the formation of water molecules.

During charge, the chemical reactions are reversed so that the active material is converted back to lead dioxide on the positive plates and to spongy lead on the negative plates.

The above reactions may be summarized thus:

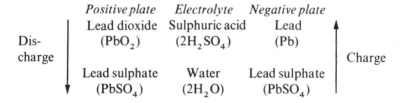

	Positive plate	*Electrolyte*	*Negative plate*	
Dis-charge	Lead dioxide (PbO_2)	Sulphuric acid ($2H_2SO_4$)	Lead (Pb)	Charge
	Lead sulphate ($PbSO_4$)	Water ($2H_2O$)	Lead sulphate ($PbSO_4$)	

It will be seen that for every two molecules of sulphuric acid decomposed during discharge, two molecules of water are formed; hence the density of the electrolyte falls as the cell discharges. The reverse process occurs during charging, so that when the cell is fully charged, the density of the electrolyte is restored to its initial value.

21.7 Alkaline cells

In both the nickel–iron and the nickel–cadmium types, the positive plates are made of nickel hydroxide enclosed in finely perforated steel tubes or pockets, the electrical resistance being reduced by the addition of flakes of pure nickel or graphite. These tubes or pockets are assembled in nickelled-steel plates. In the nickel–iron cell the negative plate is made of iron oxide with a little mercuric oxide to reduce the resistance, the mixture being enclosed in perforated steel pockets, also assembled in nickelled-steel plates. In the nickel–cadmium cell the active material is cadmium mixed with a little iron, the purpose of the latter being to prevent the active material caking and losing its porosity.

In both types of cell, the electrolyte is a solution of potassium hydroxide (KOH) having a relative density of about 1·15–1·2, depending upon the type of cell and the conditions of service. The electrolyte does not undergo any chemical change; consequently the quantity of electrolyte can be reduced to the minimum necessitated by adequate clearance between the plates.

The plates are separated by insulating rods and assembled in sheet-steel containers, the latter being mounted in non-metallic crates to insulate the cells from one another.

The advantages of the alkaline accumulator are: (*a*) its mechanical construction enables it to withstand considerable vibration, and (*b*) it is free from 'sulphating' or any similar trouble and can therefore be left in any state of charge without damage. Its disadvantages are: (*a*) its cost is greater than that of the corresponding lead cell; (*b*) its average discharge p.d. is about 1·2 V compared with 2 V for the lead cell, so that for a given voltage the number of alkaline cells is about 67 per cent greater than that of lead cells.

Due partly to these disadvantages of the alkaline cell and partly to the improvements made in the construction of the lead–acid cell —especially the portable type—during the past twenty years, the great majority of modern batteries are of the lead–acid type.

Example 21.1 *A constant terminal voltage of* 120 *V is to be maintained by a battery of alkaline cells. The initial and final values of the e.m.f. per cell are* 1·3 *V and* 1·15 *V respectively. The internal resistance per cell is* 0·01 Ω *and the discharge current is* 10 *A. Calculate the initial and final number of cells required in series.*

Initial terminal voltage/cell = initial e.m.f./cell — voltage
drop/cell due to internal resistance
= 1·3 [V] — (10 [A] × 0·01 [Ω])
= 1·3 — 0·1 = 1·2 V.

∴   initial number of cells = 120 [V]/1·2 [V] = 100.

Final terminal voltage/cell = 1·15 − 0·1 = 1·05 V,

∴   final number of cells = 120 [V]/1·05 [V] = 114.

Example 21.2 *A battery of* 30 *lead-acid cells is to be charged at a constant current of* 8 *A from a* 110-*V d.c. supply. The terminal voltage per cell is* 1·9 *V at the commencement of charging and* 2·6 *V at the end. Calculate the maximum and minimum values of the resistor required in series with the battery.*

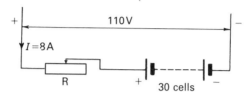

Fig. 21.6 Circuit diagram for Example 21.2.

Fig. 21.6 represents a variable resistor R connected in series with the battery of 30 cells.

At commencement of charging,

total p.d. across battery = 1·9 [V] × 30 = 57 V,

∴   corresponding p.d. across R = 110 − 57 = 53 V

and   corresponding resistance of R = 53 [V]/8 [A] = 6·625 Ω.

At end of charging,

total p.d. across battery = 2·6 [V] × 30 = 78 V,

∴   corresponding p.d. across R = 110 − 78 = 32 V,

and   corresponding resistance of R = 32 [V]/8 [A] = 4 Ω.

Example 21.3 *A battery of* 80 *cells is charged through a fixed resistor from a* 240-*V d.c. supply. At the beginning of the charge, the e.m.f. per cell is* 1·9 *V and the charging current is* 5 *A. The internal resistance per cell is* 0·06 Ω. *Calculate the value of the resistor.*

At beginning of charge,

terminal voltage/cell = e.m.f./cell + voltage drop/cell due to internal resistance

= 1·9 [V] + (5 [A] × 0·06 [Ω])

= 1·9 + 0·3 = 2·2 V.

total voltage across battery = 2·2 [V] × 80 = 176 V.

Corresponding voltage across resistor $= 240 - 176 = 64$ V,

so that value of resistor $= 64$ [V]/5 [A] $= 12\cdot8$ Ω.

Example 21.4 *A fully-charged lead-acid cell was completely dis-charged in 10 h, the discharge current being constant at 6 A. The average terminal voltage during discharge was $1\cdot95$ V. A charging current of 4 A, maintained constant for 17 h, was required to restore the cell to its initial state of charge, the average terminal voltage being $2\cdot3$ V. Calculate* (a) *the ampere hour efficiency and* (b) *the watt hour efficiency.*

The *ampere hour efficiency* of a cell is the ratio of the number of ampere hours obtainable during discharge to that required to restore the cell to its original condition.

The *watt hour efficiency* of a cell is the ratio of the number of watt hours obtainable during discharge to that required to restore the cell to its original condition.

(*a*) Output of cell, in ampere hours $= 6$ [A] $\times$ 10 [h] $= 60$ A h.

Input to cell, in ampere hours $= 4$ [A] $\times$ 17 [h] $= 68$ A h.

$\therefore$ ampere hour efficiency $= \dfrac{60 \text{ [A h]}}{68 \text{ [A h]}} = 0\cdot882$ per unit

$= 88\cdot2$ per cent.

(*b*) Output of cell, in watt hours $= 6$ [A] $\times$ $1\cdot95$ [V] $\times$ 10 [h]

$= 117$ W h.

Input to cell, in watt hours $= 4$ [A] $\times$ $2\cdot3$ [V] $\times$ 17 [h]

$= 156\cdot4$ W h.

$\therefore$ watt hour efficiency $= \dfrac{117 \text{ [W h]}}{156\cdot4 \text{ [W h]}} = 0\cdot748$ per unit

$= 74\cdot8$ per cent.

Summary of Chapter 21

In both primary and secondary cells, chemical energy is converted into electrical energy when the cells are discharging; but primary cells can only be replenished by renewal of active material, whereas in secondary cells the chemical action is reversible. Descriptions are given of the dry Leclanché and mercury cells and of the lead–acid and alkaline types of secondary cells.

EXAMPLES 21

1. An accumulator is overcharged by 5 A for 10 h. If the electrochemical equivalents of hydrogen and oxygen are 0·010 45 mg/C and 0·082 95 mg/C respectively, calculate the volume of water required to be added to compensate for gassing. Assume the mass of 1 mm³ of water to be 1 mg.

2. A current of 20 A is supplied by a cell having an e.m.f. of 2 V to a load with a terminal voltage of 1·8 V. What is the internal resistance of the cell?
(E.M.E.U., G2)

3. If the terminal voltage of a lead–acid cell varies between 2·1 V and 1·85 V during discharge, calculate the number of cells required to give 230 V (*a*) at the beginning of discharge and (*b*) at the end of discharge.

4. Calculate the number of lead–acid cells to be connected in series to give a terminal voltage of 240 V when the battery is supplying a current of 12 A. Assume each cell to have an e.m.f. of 2 V and an internal resistance of 0·025 Ω.

5. A battery of 50 cells in series is charged through a 4-Ω resistor from a 230-V supply. If the terminal voltage per cell is 2 V and 2·7 V respectively at the beginning and the end of the charge, calculate the charging current (*a*) at the beginning and (*b*) at the end of the charge.

6. A battery of 40 cells in series is to be charged from a 240-V supply. The average terminal voltage per cell during charge is 2·2 V. Calculate the value of the resistor required in series with the battery to give an average charging current of 5 A.

7. A battery of 10 cells in series is to be charged at a constant current of 8 A from a generator having an e.m.f. of 35 V and an internal resistance of 0·2 Ω. The internal resistance of each cell is 0·08 Ω and its e.m.f. ranges from 1·85 V discharged to 2·15 V charged. Calculate the maximum and minimum values of the series resistor required in the circuit.

8. A battery of 12 cells in series is charged through a fixed resistor from a 30-V supply. When charging commences, the e.m.f. per cell is 1·85 V and the charging current is 4 A. At the end of the charge, the e.m.f. per cell has risen to 2·25 V. If each cell has an internal resistance of 0·05 Ω, calculate the value of the external resistor and the current at the end of the charge.

9. An alkaline cell is discharged at a constant current of 5 A for 10 h, the average terminal voltage being 1·2 V. A charging current of 3 A, maintained for 21 h, is required to bring the cell back to the initial state of charge, the average terminal voltage being 1·48 V. Calculate the ampere hour and watt hour efficiencies.

ANSWERS TO EXAMPLES 21

1. 16 800 mm³.
2. 0·01 Ω.
3. 110 cells, 125 cells.
4. 141 cells.
5. 32·5 A, 23·75 A.

6. 30·4 Ω.
7. 1·062 Ω, 0·687 Ω.
8. 1·35 Ω, 1·54 A.
9. 79·4 per cent, 64·4 per cent.

CHAPTER 22

Electrical measurements

22.1 Current and voltage measurement

There are essentially two types of instrument used for the measurement of current and voltage, analogue and digital instruments. Analogue instruments show the quantity being measured, perhaps a current, converted into another quantity, perhaps the deflection of a pointer, which is directly related to the current. The deflection is analogous to the current. With digital instruments the reading appears in the form of a series of digits. An example of an analogue form of display is the watch where the time is indicated by the analogous movement of pointers round a scale. A watch showing the time in the form of digits is an example of a digital instrument.

The currents or voltages to be measured can be either steady quantities or alternating. Thus, for example, the current and voltage produced when a battery is connected to a simple circuit containing only resistance is steady direct current or voltage. However the current and voltage produced in a circuit connected to the mains supply is alternating current and voltage. Fig. 22.1 shows the basic form of such alternating quantities. The number of cycles per second is known as the frequency, the unit being the hertz (Hz) where 1 Hz is one cycle per second. The mains voltage has a frequency in this country of 50 Hz.

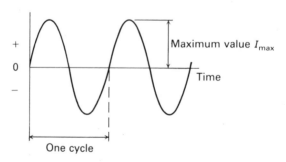

Fig. 22.1 Alternating current or voltage.

. While instruments can be used to determine the maximum value of the current or voltage it is more common to determine a quantity called the root mean square value. This is the value of the steady current that would be needed to give the same electrical power, i.e. dissipate energy at the same rate. The power of an electrical current I when passing through a resistance R is I^2R (see section 17.3). With an alternating current the power is not constant because the value of I^2 is changing. Fig. 22.2 shows how I^2 varies with time for the alternating current having the waveform shown in fig. 22.1. The average value of

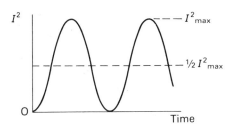

Fig. 22.2

the I^2 graph, over a complete cycle, is $\frac{1}{2}I^2_{max}$. Thus the equivalent steady current must be such that

$$(\text{equivalent current})^2 = \tfrac{1}{2}I^2_{max}$$

$$\text{equivalent current} = \frac{I_{max}}{\sqrt{2}}$$

This effective current is called the root mean square current, because it is the square root of the mean or average values of the squares of the current.

$$\text{root mean square current} = \frac{I_{max}}{\sqrt{2}} \qquad (22.1)$$

Similarly

$$\text{root mean square voltage} = \frac{V_{max}}{\sqrt{2}} \qquad (22.2)$$

Example 22.1 *The mains voltage has a root mean square voltage of 240 V, what is the maximum value of this voltage?*

Since root mean square voltage $= \dfrac{V_{max}}{\sqrt{2}}$

$$V_{max} = \sqrt{2} \times 240\,[V]$$
$$= 339\cdot4\,V$$

22.2 The moving-coil instrument

Section 19.11 of this book gives details of the construction of the moving-coil meter and the principles behind its operation. As the section indicates, the instrument can be used both as an ammeter and the basis of a voltmeter. It can also be developed into a multi-range ammeter or voltmeter by the use of appropriate shunts and multipliers. The instrument is very widely used.

The basic instrument is a steady current measuring instrument and cannot without modification be used for alternating currents, or voltages. The modification that permits the moving-coil instrument to be used with alternating current is to use it in conjunction with a rectifier circuit, this converting the alternating waveform into a unidirectional current or voltage. When such a circuit is used the instrument gives root mean square values. Such instruments can be used for frequencies up to about a few kilohertz.

Another modification of the instrument also enables it to be used for the measurement of resistance, the resulting arrangement being known as an ohmmeter. Fig. 22.3 shows the basic circuit used. When the terminals are short-circuited, i.e. there is zero resistance being measured, the variable resistor is adjusted until there is a full-scale deflection of the meter. This deflection then corresponds to zero resistance. When a resistor is inserted between the terminals the total resistance in the circuit is increased and hence the current registered by the meter decreases. The amount by which it decreases is a

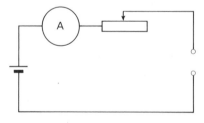

Fig. 22.3 The Ohmmeter circuit.

measure of the resistance. Hence the scale of the moving-coil meter can be given a resistance scale, starting from a zero at the maximum current end.

22.3 The cathode-ray oscilloscope

The output from a moving-coil instrument appears, generally, as the movement of a metal pointer across a scale. The output from a cathode-ray oscilloscope (fig. 22.4) appears as the movement of a spot of light across a fluorescent screen, the spot being produced by the impact of a beam of electrons on a fluorescent screen. The spot can be made to move both vertically, known as the Y-direction, and horizontally, known as the X-direction. If a voltage is connected to the Y-input terminals of the instrument the spot will move vertically, the

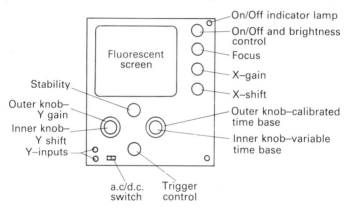

Fig. 22.4 A basic form of cathode-ray oscilloscope.

amount it moves being directly proportional to the applied voltage. If a voltage is applied to the X-input terminals the spot will move horizontally, the amount moved being proportional to the applied voltage.

For many applications the X-input is supplied from a varying voltage within the instrument. This voltage varies in such a way that the spot moves in a regular fashion across the screen so that its position in the horizontal direction across the screen is directly proportional to time. The spot moves from left to right and when it reaches the end of its travel is made to whip back very quickly to the left to start its regular motion all over again. This internal X-input is known as the time base. The speed at which the spot moves across the screen can be chosen by selecting the appropriate time base. These are expressed in terms of time per centimetre of movement in the X-direction, e.g. 10 ms/cm means that the spot will take 10 milli-seconds to cover 1 cm.

The input to the Y-terminals is used to give a deflection of the spot in the vertical direction. The amount of movement in this direction can be, for a given input voltage, varied by choosing different amplification factors for the input. Thus the Y-input amplifier could be chosen to give 10 V/cm. This would mean that for every 10 V of input the spot would move 1 cm in the vertical direction.

If there is no time base switched on, and no input to the X-input terminals, then when a d.c. voltage is applied to the Y-input terminals the spot will be displaced in a vertical deflection from its zero position to a new position corresponding to the applied voltage. The amount of this displacement can be measured and, from a knowledge of the Y-amplification factor or by calibrating with known d.c. voltages, the voltage calculated.

If the time base is switched on when a d.c. voltage is applied to the Y-input terminals, then there will be initially a straight line across the screen due to the moving spot and then the input will cause this entire line to be displaced. The amount of this displacement can be measured and the voltage calculated as before.

A more important use of the cathode-ray oscilloscope is, however, to display the waveform of an alternating voltage. With the time base switched on the alternating voltage is applied to the Y-input terminals. The spot in moving across the screen will then trace out the form of the input voltage. Without any further adjustments it is likely that the trace on the screen will be moving and not stationary. This is because when the spot reaches the end of its path and whips back to the beginning again it is unlikely to start at exactly the same point on the waveform it was on in previous movement across the screen. It can be made to start at the same point on the waveform every time by adjusting a control called the trigger. This ensures that the spot will only trigger its movement across the screen when a particular voltage is reached. The result is a stationary trace on the screen. Measurements can be made on the trace, e.g. the maximum value of the voltage can be directly measured or the time taken to complete a cycle determined and so the frequency of the waveform calculated.

Example 22.2 *Fig. 22.5 shows the trace on a cathode-ray oscilloscope screen when the time base is 1 ms/cm and the Y-amplification is 5 V/cm.*

What is (a) *the frequency of the waveform and* (b) *the maximum voltage?* (Each square on the screen has a side of 1 cm.)

(a) One complete cycle occupies 4 cm and so the time taken to complete one cycle is $4 \times 1 = 4$ ms. The frequency is the number of cycles per second and so

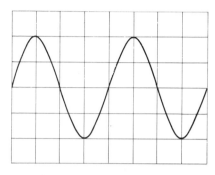

Fig. 22.5

$$\text{frequency} = \frac{1}{4}[\text{ms}]$$

$$= \frac{1}{4 \times 10^{-3}[\text{s}]}$$

$$= 250 \text{ Hz}.$$

(b) The maximum voltage involves a displacement in the Y-direction of 2 cm. Hence

$$V_{\text{max}} = 2 \times 5 = 10 \text{ V}.$$

22.4 Digital voltmeters

These instruments take the voltage to be measured and convert it into a series of digits. One method that is used involves counting the number of pulses from an accurate electronic clock that occur during the time taken for a steadily rising voltage to rise from zero until it reaches the value of the voltage being measured. The pulse count is then a measure of the voltage.

Summary of Chapter 22

With an analogue instrument the quantity being measured is converted into an analogous quantity of some other variable. With a digital instrument the quantity is converted into a series of digits.

$$\text{Root mean square current} = \frac{I_{max}}{\sqrt{2}} \qquad (22.1)$$

$$\text{Root mean square voltage} = \frac{V_{max}}{\sqrt{2}} \qquad (22.2)$$

One of the most common instruments for the measurement of steady currents and voltages is the moving-coil instrument, and with a rectifier circuit also alternating currents and voltages. The cathode-ray oscilloscope can be used for the measurement of steady and alternating voltages. It also enables the waveform of a fluctuating voltage to be seen.

EXAMPLES 22

1. What is the maximum value of a current taken from the mains supply when it has a root mean square value of 2·0 A?
2. What is the maximum value of the mains voltage if it is 240 V root mean square?
3. A cathode-ray oscilloscope is used to measure the maximum voltage of a sinusoidal waveform and a value of 12 V is obtained. What is the root mean square value of this voltage?
4. Explain the difference between an analogue instrument and a digital instrument.
5. Explain how the cathode-ray oscilloscope can be used to determine the waveform of an alternating voltage.
6. A repetitive waveform is found by means of a cathode-ray oscilloscope to have a time of 1·0 ms between successive positive maxima. What is the frequency of the waveform?

ANSWERS TO EXAMPLES 22

1. 2·8 A
2. 339 V
3. 8·5 V
6. 1000 Hz

CHAPTER 23

Wave Motion

23.1 Transverse wave motion

One of the commonest examples of wave motion is produced when a
stone is thrown into a pool of perfectly still water. The ripple travels
radially outwards but the movement of a floating object indicates that
the water at any particular point merely moves up and down. Thus,
while the crest of the wave in fig. 23.1 travels from A to B, the

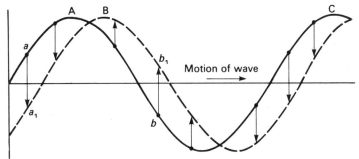

Fig. 23.1 Transverse wave motion.

particles of water at a move down to a_1 while those at b move up to b_1,
etc.

The distance between two adjacent crests A and C is termed the
wavelength and is represented by the Greek letter λ (lambda); and
the time taken for the wave to travel one wavelength is termed the
periodic time and is represented by T. If v is the speed at which the
wave travels outwards, then:

$$v = \frac{\text{one wavelength}}{\text{time to travel one wavelength}}$$
$$= \lambda/T \tag{23.1}$$

If f represents the number of *crests* passing a given point, say C,
per second, the *frequency* of the wave is said to be f *cycles/second* or
hertz (see section 20.8), then:

$$f = 1/T$$
$$\text{and } v = \lambda/T = \lambda f \tag{23.2}$$

The type of wave motion where the direction of movement of the particles is *perpendicular* to the direction of movement of the ripple is referred to as *transverse wave motion*.

Transverse wave motion can be set up in a long vertical rope, suspended at its upper end, by merely giving the lower end a sideways jerk. The wave travels up along the rope to the point of support and is then reflected back to the lower end. Light and radio are transverse wave motions.

23.2 Longitudinal wave motion

With a longitudinal wave motion the direction of movement of the particles is back-and-forth along the same direction as that of the wave motion. Fig. 23.2 shows apparatus by which such a wave mo-

Fig. 23.2 A model for a longitudinal wave motion.

tion can be demonstrated. It consists of a number of trolleys linked together by springs. When the first trolley is given a back-and-forth push and pull a disturbance travels along the line, each trolley moving back-and-forth along the direction of motion of the disturbance. The effect of such motions is that, at some instant, some trolleys will have moved closer together and some further apart, as illustrated by fig. 23.3. We can talk of the wave motion being propagated as a series of compressions and rarefactions. If we plot a graph of the displacement of the trolleys along the line against their distance along the line then

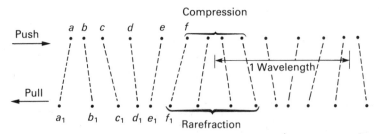

Fig. 23.3 Regions of compression and rarefaction in a longitudinal wave, the letters indicating the positions of trolley in Fig. 23.2.

a graph of the form shown in fig. 23.4 is produced and we can talk of a wavelength in the same way as we do for transverse waves.

Sound is an example of a longitudinal wave motion. It travels through a medium as a series of compressions and rarefactions. Thus, for example, a loudspeaker cone produces a sound wave in air by the electrical signal to the loudspeaker causing the speaker cone to move back-and-forth and so push the air in front of it back-and-forth.

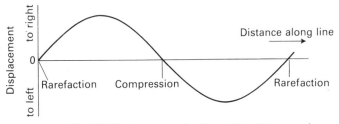

Fig. 23.4 Displacements of trolleys in Fig. 23.2.

23.3 Characteristics of sound waves

(*a*) *Sound requires a material medium.* For instance, if an electric bell is suspended and set ringing in an air-tight vessel, the sound decreases as the air is pumped out, until it becomes practically inaudible. Readmission of air enables the bell to be heard again.

The medium is necessary if sound, a longitudinal wave, is to be transmitted as a series of compressions and rarefactions.

(*b*) *Sound takes time to travel.* The speed of a sound wave in air can be measured by observing the time between the puff of smoke from a gun fired a known distance away and the instant when the sound is heard. Another example of the time taken by a sound to travel in open air is the interval between a lightning flash and the arrival of the thunder. Since the speed of sound in open air is about 0·33 km/s, the distance, in kilometres, of the lightning from the observer can be estimated by dividing the length of the interval, in seconds, by three.

The following table gives the speed of sound in some well-known media:

Medium	Temperature [°C]	Speed [km/s]
Open air (no wind)	0	0·33
Hydrogen	0	1·27
Sea water	20	1·54
Copper	20	3·56
Iron	20	5·00

The speed of sound in a medium depends on the temperature.

(c) *Sound can be reflected.* An echo is a well-known example of sound reflection and occurs when a sound wave comes up against a hard surface such as metal or stone. Another well-known example of sound reflection is the circular Whispering Gallery in the Dome of St Paul's Cathedral. A low whisper uttered *near* one side of the gallery is reflected by the curved wall and is perfectly audible at the other side of the gallery.

Echo-sounding equipment can be fitted to a ship for charting the depth of the sea. The principle involved is to transmit a sound impulse vertically downward from a diaphragm fitted below water surface on one side of the hull and measure the interval taken for the sound to travel to the bottom of the sea and then be reflected back to a receiving device at the same level on the other side of the hull.

(d) *Sound can be refracted.* It is well-known that sounds such as those of church bells, trains, etc., can be heard when the wind is from their direction, though they may be quite inaudible at other times.

If the direction of the wind is the same as that in which the sound is travelling, the velocity of the sound relative to earth is the sum of the wind and sound velocities. The effect may be complicated by the fact that the velocity of the wind is usually greater in the upper air layers than near the ground, so that the higher the air layer, the greater, in general, is the sound velocity relative to earth. This condition gives refraction of a sound wave i.e. the direction of travel of the wave changes.

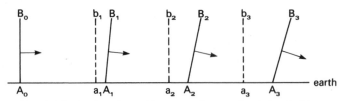

Fig. 23.5 Refraction of sound by cumulative sound and air velocities relative to earth.

In fig. 23.5, A_0B_0 represents an end elevation of a wavefront of a sound wave moving from left to right. When the air is *perfectly still*, the positions of the wavefront after *equal* intervals of time are represented by the vertical dotted lines, a_1b_1, a_2b_2, etc.

If the wind is blowing in the *same* direction as the sound wave, and if it travels distance a_1A_1 near the ground and distance b_1B_1 in an upper layer during the time it takes the sound wave to travel distance A_0a_1 in still air, the new wavefront is tilted *forward* as represented by A_1B_1. Consequently, the sound wave is concentrated more and more along the surface of the earth, as represented by A_2B_2 and A_3B_3.

Figure 23.7 shows the effect of the wind blowing in opposition to the sound wave. The net velocity of the sound in the upper layers

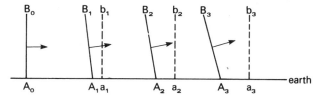

Fig. 23.6 Refraction of sound by differential sound and air velocities relative to earth.

is now less than that near the ground, with the result the wavefront is tilted upwards and the sound tends to be dissipated in the upper air layers.

Another reason for the possible refraction of a sound wave transmitted by air is the variation of the temperature at different altitudes—the higher the air temperature, the greater the speed of sound transmission. If the air temperature were uniform, the speed of sound would also be uniform—any variation in the air pressure does not affect the speed of sound. Consequently, in the absence of wind, the wavefronts in figs. 23.5 and 6 would be represented by a_1b_1, a_2b_2, etc. If, however, the temperature increases with height, as often happens about sunset, the speed of sound also increases with height, and the wavefront tilts towards the ground, as shown in fig. 23.5, and the sound will therefore be heard more clearly at a distance. The converse effect, shown in fig. 23.6, occurs when the temperature falls with height above ground, as is usually the case during daytime.

Any change in the velocity of sound due to variation in the composition of the transmitting medium causes the direction of motion of the wavefront to be deflected, and the effect is known as *refraction*—an effect that is similar to the deflection of a beam of light when it passes from, say, air to another medium such as glass or water.

23.4 Characteristics of the sensations of sound

The sensation of sound is received via our ears and depends upon three characteristics, namely:

(*a*) *Pitch*. The pitch of a sound in music or speech depends upon the frequency of the vibrations—the higher the frequency, the higher is the pitch. For instance, the middle C of a piano has a frequency of 256 hertz, if tuned to the Scientific Scale. If the frequency is doubled to 512 hertz, the pitch is increased by an octave. On the other hand, if the frequency is halved to 128 hertz, the pitch is lowered an octave. In other words, the pitch of a note is its position on a musical scale and rises with increase of frequency.

The lowest frequency that can be detected as a musical note by the human ear is about 30 hertz and the upper limit is about 15 kilo-hertz, but this limit decreases with age.

b) *Loudness*. The louder the sound from a given source, the greater the to-and-fro movement of the particles of the transmitting medium. A variation in the loudness of a note is not accompanied by any change of pitch. The power required for a note to be just audible varies considerably at different frequencies, the human ear being most sensitive for sounds having a frequency of about 2 kHz.

(*c*) *Quality*. The quality of a sound is that which distinguishes it from another sound of the *same pitch* emitted by a different source. For instance, notes of the *same* pitch played on, say, a violin and a trumpet sound quite different. The distinctive quality of a musical note is due to the presence of other pitch notes of various amplitude which accompany the principal note called the *fundamental*.

Summary of chapter 23

A transverse wave motion has the direction of movement of particles perpendicular to the direction of motion of the wave, a longitudinal wave motion has particle movement along the same line as that of the wave motion.

$$v = \lambda/T = \lambda f \qquad (23.2)$$

EXAMPLES 23

1. In an experiment on transverse waves in a large tank nearly filled with water, the ripples travelled 40 cm in 2·3 s, and had a wavelength of 2·5 cm. Calculate (*a*) the speed of the ripples in centimetres/second, (*b*) the frequency of the ripples.

2. A person sees a flash of lightning 12 s before he hears the first sound of thunder. Neglecting the effect of any wind, calculate the distance of the lightning from the observer. Assume the velocity of sound in the open air to be 0·33 km/s.

3. How long does it take a sound to travel 2 km in calm air? If there is a wind blowing at 50 km/h, calculate the time for the sound to travel the 2 km when the wind is (*a*) in the same direction as that of the sound and (*b*) in direct opposition.

4. A tuning fork has a frequency of 256 Hz. Calculate the wavelength of the note, assuming the speed of sound to be 0·33 km/s. Also calculate the wavelength of a note that is an octave above that emitted by the fork.

5. A disc has 90 holes arranged in a circle. If the disc is driven at 500 rev/min and air is forced from a nozzle through each of the holes in turn, what is the frequency of the fundamental note produced?

6. The lowest frequency that is audible is about 30 Hz and the highest is about 15 kHz. What is the wavelength in each case? What is approximately the number of octaves in this range? Assume the velocity of sound in air to be 0·33 km/s.

7. With an echo depth-sounding equipment, it was found that the echo was received 1·2 s after the sound impulse was transmitted. Calculate the depth of the sea-bed below the equipment. Assume the speed of sound in sea water to be 1·54 km/s.

8. A steamer approaching a high cliff gives a short whistle and the echo is received 8 s later. The whistle is repeated after 5 minutes and the interval for the echo is found to have decreased to 4 s. Assuming the ship to be travelling at a constant speed, calculate the value of this speed in kilometres/hour.

ANSWERS TO EXAMPLES 23

1. 17·4 cm/s; 6·96 Hz.
2. 3·96 km.
3. 6·06 s; 5·8 s, 6·33 s.
4. 1·29 m; 0·645 m.

5. 750 Hz.
6. 11 m; 2·2 cm; 9 approx.
7. 0·924 km.
8. 7·92 km/h.

Index

Index